Lecture Notes in Mathematics

Edited by A. Dold, B. Eckmann and F. Takens

T0222650

1449

Włodzimierz Odyniec
Grzegorz Lewicki

Minimal Projections in Banach Spaces

Problems of Existence and Uniqueness and their Application

Springer-Verlag
Berlin Heidelberg New York London
Paris Tokyo Hong Kong Barcelona

Authors

Włodzimierz Odyniec
Pedagogical University
Nab reki Mojki 48
Leningrad 191186, USSR

Grzegorz Lewicki
Jagiellonian University, Department of Mathematics
Reymona 4, 30-059 Krakow, Poland

Mathematics Subject Classification (1980): 41A35, 41A52, 41A65, 46B99, 47A30, 47B05, 47B38

ISBN 3-540-53197-1 Springer-Verlag Berlin Heidelberg New York
ISBN 0-387-53197-1 Springer-Verlag New York Berlin Heidelberg

Printing and binding: Druckhaus Beltz, Hemsbach/Bergstr.
2146/3140-543210 – Printed on acid-free paper

To

Marii Schiffmann (Elenbogen)

Preface

This book results in part from lectures held by the first author at the Pedagogical University of Bydgoszcz in 1984 - 85. A part of the present text (chapters I, II, IV) is a revised version of lecture notes [156], which appeared in 1985 under the same title (in Russian) in 350 copies, in the form of mimeographed notes (a university issue). Chapter III includes new results.

The text is supplemented with an Author Index and Subject Index; each chapter is followed by short comments on the history of results exposed in the chapter.

It is the authors' hope that the book may be of interest to those who are just beginining to deepen their knowledge of Banach space geometry and approximation theory, as well as to specialists. The second chapter can perhaps prove to be interesting also to specialists in mathematical economy.

This book would never have appeared without encouraging help from many persons.

First of all, the authors are grateful to Professor Ward Cheney for his suggestion to submit the manuscript to this series and for his interest and encouragement along the way to its completion.

Also our thanks go to Professors Yu. D. Burago, M. I. Kadeč, S. A. Konjagin, A. Pełczyński, S. Rolewicz, P. Wojtaszczyk, W. Wojtyński, W. A. Zalgaller, Dr. V. M. Kadeč, Dr. A. Koldobskii and Dr. K. Kürsten for fruitful discussions about problems dealt with in this book.

Special gratitude is due to Professor Czesław Bessaga who has read through the first version of the text and helped a lot in elaborating the final version.

Finally, we are glad to be able to thank Dr. M. E. Kuczma, Mr M. Wójcikiewicz and Miss J. Wójcikiewicz who translated the text into English.

The authors also would like to express their gratitude to Dr. M. Baran for his help in preparing the final version of this book on computer.

Włodzimierz Odyniec and Grzegorz Lewicki

Kraków, April 1990

Table of Contents

Introduction

§ 1. General survey

The first 30 years of our century , the time when the edifice of functional analysis was raised, is also the period during which the foundations of the theory of best approximations in normed space were laid. The name forever connected with the origins of that theory is that of S. Banach's ([7]).Later on the ideas of Banach were developed in the works of S. Mazur, M. G. Krein, S. M. Nikolskii, N. I. Ahiezer, J. Walsh, A. N. Kolmogorov, A. I. Markusevič, R. James, R. Phelps, A. L. Garkavi, I. Singer, E. W. Cheney, S. B. Stečkin, and others (see [72],[124],[183],[192] for detailed references).

The problem of best approximation in a normed linear space X by elements of a subspace D (here and elsewhere, by a subspace we always mean a closed subspace) is formulated as follows. Given $x \in X$, find an element $y_x \in D$ for which the lower bound

$$\rho(x,D) = \inf\{\|x - y\|: y \in D\} \qquad (0.1.1)$$

is attained.

The number $\rho(x,D)$ is called the best approximation, or a distance, to x from D. Element y_x (which need not be unique) is called an element of best approximation from D to x.

The question of vital importance in computing (or estimating) the quantity $\rho(x,D)$, as well as in veryfying is whatever or not the subspace D is complemented in X; that is to say, whether there exists a projection of X onto D. In particular, if P is such a projection, then we have for every $x \in X$

$$\|x - Px\| \le \|I - P\|\cdot\rho(x,D), \qquad (0.1.2)$$

with I denoting the identity operator in X. Since

$$\|I - P\| \le 1 + \|P\|, \qquad (0.1.3)$$

with equality

$$\|I - P\| = 1 + \|P\| \qquad (0.1.4)$$

holding in many important specific cases (see e.g. I. K. Daugavet's paper [54]), the problem which naturally arises is that of estimating fron below the norm of P. To serve this purpose, a quantity $\rho(X,D)$ is introduced, called the relative projection constant:

$$\rho(X,D) = \inf\{\|P\|: P \text{ is a projection of X onto D}\} \qquad (0.1.5)$$

(see [25],[78],[88],[90]).

Among all projections from X onto D, those are of special interest whose norms coincide with the constant $\rho(X,D)$; they are called **minimal projections** (provided they exist). In other words, a projection P of a Banach space X onto a subspace D is minimal if

$$\|P_1\| \geq \|P\| \qquad (0.1.6)$$

for any other projection P_1: $X \to D$.

The problem of finding a best approximating element, as well as that of finding a minimal projection, is in fact an instance of an optimization problem. The relationship between the two types of problems is not just apparent; it goes quite deep into the inner nature of things, though certain differences occur at some points. Similarly to the general task of finding a best approximation, when dealing with minimal projections we are facing two principal kinds of problems:

(E_m) = problem of existence of minimal projection,

(U_m) = problem of uniqueness of a minimal projection.

In a finite-dimensional space, problem (E_m), just as (E), the problem of existence of a nearest element, always has a solution. This is no more the case for spaces of infinite dimension (see [39],[83]). Both of this existence problems can lack a solution. Let us examine two examples. In each of them X is taken to be c_o ; for D we take two subspaces of codimension 1 defined by two functionals f_1, $f_2 \in (c_o)^* = l_1$.

Example 0.1.1. a) $f_1 = (1/2,1/4,1/8,\ldots,1/2^n,\ldots)$, $x = (2,0,\ldots) \in c_o$, $D = f^{-1}(0)$. It is easy to see that $\inf\{\|x - y\|: y \in D\} = 1$, whereas $\|x - y\| > 1$ for any particular y in D. Strangely enough, there exists a minimal projection onto D, of norm 1, which is unique (moreover it is a SUBA projection) (see [12],[21]).

b) $f_2 = (1/4,1/4,1/4,1/8,1/16,\ldots 1/2^{n+1},\ldots)$, $D = f_2^{-1}(0)$

Then $1 < \rho(X,D) < 2$ and there are no minimal projections onto D (see [21]).

To emphasize the relevance of minimal projections to the problem of best approximation we mention the following result of P. Franck ([68]). Let $\mathcal{L}(X,Y)$ denote the space of all linear operators from X to Y (Banach spaces) and let $\mathcal{L}_V$ be the space of all operators in $\mathcal{L}(X,Y)$ whose kernels contain a given subspace $V \subset X$. Then for any $T \in \mathcal{L}(X,Y)$ we have the inequality

$$\inf\{\|T - T_1\|: T_1 \in \mathcal{L}_V\} \geq \sup\{\|Tx\|_Y/\|x\|_X: x \in V\} \qquad (0.1.7)$$

which turns into an equality whenever there exists a norm one projection of X onto V. (For functionals (0.1.7) holds with an equality sign ([27], [71]).)

Despite the fact that, historically, the term "minimal projection" entered the stage relatively late, the study of minimal projections originated in fact already in the thirties, mainly in the connection with geometry of Banach spaces. Projections with unit norm have been treated most thoroughly; this is chiefly due to the fact that they can be considered as a generalization of orthogonal projections in Hilbert space.

Numerous mathematicians have been dealing with norm one projections; let us mention there A. E. Taylor, H. F. Bohnenblust, L. V. Kantorovich, R. C. James, L. Nachbin, G. P. Akilov, M. Z. Solomjak, J. Lindenstrauss, M. I. Kadeč, Cz. Bessaga, A. Pełczyński, J. Ando, V. I. Gurarii (see[19], [139], [159] for detailed references).

On the other hand, papers dealing with certain minimal projections with nonunit norm appeared already in the forties. We wish to emphasize expecially two of these works, A. Sobczyk [180], 1941, and S. M. Łozinskii [130], 1948.

Using the results of Taylor [186] and Bohnenblust [25], A. Sobczyk [180] has shown that minimal projections of (c) onto (c_o) have norm 2 and there are infinitely many of them. (Recall that (c_o) is of codimension 1 in (c)). To be precise, if $(x) \in (c)$ then, writing $x_o = (1,1,\dots)$, we can find a unique $t \in \mathbb{R}$ such that $x = x^o + t\cdot x_o$.

Now, if we define an operator P by

$$Px = x + t\cdot b, \tag{0.1.8}$$

where

$$b = (b_1, b_2, \dots) \in (c), \quad \lim_{i\to+\infty} b_i = -1 \tag{0.1.9}$$

then clearly P is a projection of (c) onto (c_o); if besides $|b_i| \le 1$ for all i, then P is a minimal projection.

Łozinskii's paper [130] is connected with the Fourier projection F_n of $C_o(2\pi)$, the space of all continuous, 2π-periodic function, onto π_n, the space of all trigonometric polynomials of order $\le n$ $(n \ge 1)$. It is defined by the formula

$$F_n(f) = \sum_{k=0}^{2n} \left(\int_0^{2\pi} f(t)\cdot g_n(t)\, dt\right) g_k, \tag{0.1.10}$$

where $(g_k)_{k=0}^{2n}$ is any orthonormal basis in π_n, i.e.

$$\int_0^{2\pi} g_i(t)\cdot g_j(t)\, dt = \delta_{ij}, \tag{0.1.11}$$

(the Kronecker delta). The main result of this paper is that $\rho(C_o(2\pi),\pi_n) = \|F_n\|$; in other words that F_n is a minimal projection (see Theorem 0.1.3

below). Because of the great importance of the Fourier projection F_n we present after [35] some basic properties of this operator. We start with the following

Theorem 0.1.2. (see [35], p. 212) The norm of F_n satisfies the following estimation:

$$4/\pi^2 \cdot \ln(n) \le \|F_n\| \le \ln(n) + 3 \tag{0.1.12}$$

Proof. It is well known (see e.g. [2], p. 180) that

$$\|F_n\| = (2/\pi) \cdot \int_0^{\pi} |D_n(t)|\, dt, \tag{0.1.13}$$

where

$$D_n(t) = \sin((n + 1/2)\cdot t)/\sin((1/2)\cdot t) \tag{0.1.14}$$

for $t \in [0,2\pi]$. The change of variable $t \to 2x$ and the inequality $\sin(x) \le x$ on $[0,2\pi]$ give us, following (0.1.13)

$$\|F_n\| \ge (2/\pi) \cdot \int_0^{\pi/2} |\sin((2n+1)x)/x|\, dx.$$

The change of variable $x \to \pi \cdot x/(2n+1)$ then yields

$$\|F_n\| \ge (2/\pi) \cdot \int_0^{n+1/2} |\sin(\pi x)|/x\, dx > (2/\pi) \cdot \int_0^{2n} |\sin(\pi x)|/x\, dx.$$

Breaking the interval into n subintervals, we obtain

$$\|F_n\| \ge (2/\pi) \cdot \left(\int_0^1 + \int_1^2 + \ldots + \int_{n-1}^n \right) |\sin(\pi x)|/x\, dx =$$

$$= (2/\pi) \cdot \int_0^1 (1/x + 1/(x+1) + \ldots + 1/(x+n-1)) \cdot |\sin(\pi x)|\, dx \ge$$

$$\ge (2/\pi) \cdot \int_0^1 (1 + 1/2 + \ldots 1/n) \cdot \sin(\pi x)\, dx \ge$$

$$\ge (2/\pi) \cdot \ln(n) \cdot \int_0^1 \sin(\pi x)\, dx = (4/\pi^2) \cdot \ln(n).$$

That $1 + 1/2 + \ldots + 1/n > \ln(n+1)$ may be seen from a graph of the function $1/(x+1)$ or by induction.

To prove the second inequality, first observe that

$$D_n(t) = 1/2 + \sum_{i=1}^{n} \cos(it) \tag{0.1.15}$$

Integrating separately over $[0,1/n]$ and $[1/n,\pi]$ we obtain the following estimations

$$(2/\pi) \cdot \int_0^{1/n} |D_n(t)|\, dt = (2/\pi) \cdot \int_0^{1/n} 1/2 + \sum_{i=1}^{n} |\cos(it)|\, dt \le$$

$$\le (2/\pi \cdot n) \cdot (1/2 + n) < 1$$

and

$$(1/\pi) \cdot \int_{1/n}^{\pi} |D_n(t)|\, dt \le (1/\pi) \cdot \int_{1/n}^{\pi} \pi/t\, dt = \ln(\pi) - \ln(1/n) < 2 + \ln(n)$$

(here we have used the fact that $\sin(t/2) \ge t/\pi$ which is evident from the graph of $\sin(t/2)$. Combining the above inequalities we derive the second estimation. ***

Theorem 0.1.3 ([35],p.212) Let P be any projection of the space $C_o(2\pi)$ onto the space π_n. Then $\|P\| \geq \|F_n\|$. In other words F_n is a minimal projection among all projections of $C_o(2\pi)$ onto π_n.

Proof. Define operators T_s and Φ by the equations

$$(T_s f)(x) = f(x+s)$$

$$(\Phi f)(x) = (1/2\pi) \cdot \int_{-\pi}^{\pi} (T_{-s} P T_s f)(x)\, ds.$$

If we can establish that $\Phi = F_n$, then we will be finished because

$$\|F_n f\| = \|\Phi f\| = \max(|(1/2\pi) \cdot \int_{-\pi}^{\pi} (T_{-s} P T_s f)(x)\, ds : x \in [0,2\pi]) \leq$$

$$\leq \|T_{-s} P T_s f\| \leq \|P\| \cdot \|f\|.$$

In order to prove that $F_n = \Phi$ it will be enough to prove that $\Phi f_k = F_n f_k$ where f_k is the function $f_k(x) = e^{ikx}$ $(k=0,\pm1,\pm2,\ldots)$, since this family of functions is fundamental in $C_o(2\pi)$, while the operators Φ and F_n are linear and continuous. If $|k| \leq n$, then $F_n f_k = f_k$. On the other hand, $T_s f_k \in \pi_n$, so that $P T_s f_k = T_s f_k$. Thus $T_{-s} P T_s f_k = f_k$ and $\Phi f_k \quad f_k$. (In the integration, the integrand is independent of s.) Suppose next that $|k| > n$. Then, by (0.1.10) and (0.1.11), $F_n f_k = 0$. Since $T_s f_k = e^{iks} \cdot f_k$, it follows that $(T_{-s} P T_s f_k)(x) = e^{iks} \cdot (P f_k)(x-s)$. But $P f_k \in \pi_n$, and consequently, as a function of s, e^{iks} is orthogonal to $(P f_k)(x-s)$. Hence $\Phi f_k = 0$, which completes the proof. ***

Theorem 0.1.4. ([35],p.214) Let P be a projection of the even part of $C_o(2\pi)$ onto the even part of π_n. Then $\|I - P\| \geq (1/2) \cdot (\|F_n\| + 1)$.

Proof. Define the linear operator

$$(\Phi f)(x) = (1/2\pi) \cdot \int_{-\pi}^{\pi} (T_s (I - P)(T_{-s} + T_s) f)(x)\, ds$$

in which T_s, as in the previous theorem, denotes the translation operator. The crux of the proof is in verifying the equation $\Phi = I - F_n$. After that we write $\|(I - F_n)(f)\| = \|\Phi(f)\| \leq 2 \cdot \|I - P\| \cdot \|f\|$ whence $2 \cdot \|I - P\| \geq \|I - F_n\| = 1 + \|F_n\|$. (The last equality can be obtained using Daugaviet's Teorem from [54].) In order to prove that $\Phi = I - F_n$, it sufficies to prove that $\Phi f_k = (I - F_n) f_k$, where $f_k(x) = \cos(kx)$, since these functions form a fundamental set and the operators in questions are continuous. Thus we must show that $\Phi f_k = 0$ when $k \le n$ and $\Phi f_k = f_k$ for $k > n$. We have

$$(T_s f_k)(x) = \cos(k(x+s)) = \cos(ks)\cdot\cos(kx) - \sin(ks)\cdot\sin(kx).$$

Hence $(T_{-s} + T_s)f_k = 2\cdot\cos(ks)\cdot f_k$

and $T_s(I - P)(T_{-s} + T_s)f_k = 2\cdot\cos(ks)\cdot T_s(f_k - Pf_k)$.

Now if $k \le n$, $f_k = Pf_k$ so that $\Phi f_k = 0$. If $k > n$, then

$$(\Phi f_k)(x) = (1/\pi) \int_{-\pi}^{\pi} \cos(ks)[\cos(ks)\cos(kx) - \sin(ks)\sin(kx) - (Pf_k(x+s)]ds$$

Since $(Pf_k)(x-s)$ is a trigonometric polynomial of degree $\le n$ in s, the integral involving it vanishes, by orthogonality. By the orthonormality relations the remaining integration yields $\cos(kx) = f_k(x)$. ***

The Banach-Steinhaus Theorem, Theorems (0.1.2) and (0.1.3) yield the following

Theorem 0.1.5. (Charsziładze-Łozinski Theorem I) For each n let there be given a continuous projection P_n of $C_o(2\pi)$ onto π_n. Then there exists a function $f \in C_o(2\pi)$ for which $\|P_n f\|$ is unbounded as $n \to \infty$.

Proof. By the Theorems (0.1.3) and (0.1.4) $\|P_n\|$ is unbounded because

$$\|P_n\| \ge \|F_n\| \ge (4/\pi^2)\cdot\ln(n).$$

If $\|P_n f\|$ were bounded for all f, then by the Banach-Steinhaus Theorem, $\|P_n\|$ would be bounded. ***

A simple consequence of Theorem (0.1.4) is

Theorem 0.1.6. (Charsziładze-Łozinski Theorem II) For each n, let P_n denote a projection of $C_R[a,b]$ onto the subspace of algebraic polynomials of degree $\le n$. Then there exists a function $f \in C_R[a,b]$ for which the sequence $\|f - P_n f\|$ is unbounded.

Proof. We define a map M from $C_R[a,b]$ onto the evan part of $C_o(2\pi)$ by

$$(Mf)(\alpha) = f((a+b)/2 + ((b-a)/2)\cdot\cos(\alpha)).$$

The map M is an isometric isomorphism. That is, it is one-to-one, linear, and has the property $\|Mf\| = \|f\|$. Now the operators $P_n^* = MP_nM^{-1}$ are projections of the even part of $C_o(2\pi)$ onto the evan trigonometric polynomials of degree $\le n$. By Theorem (0.1.5), then, $\|I - P_n^*\| \to \infty$. Hence $\|I - P_n\| \to \infty$, and by the Banach-Steinhaus Theorem, $\|f - P_n f\|$ is unbounded for some f. ***

The question of uniqueness of F_n has remained unsolved until 1968 (see [36],[37],and also [89]). The more general situation has been considered in ([172]).

The papers of Sobczyk [180] and Łozinski [130], although bearing very little resemblance to one another, both lie in the two main streams of application of minimal projections: geometry of Banach spaces and theory of best approximations. These two principial topics have become the object of the first general survey by E. W. Cheney and K. H. Price in 1970 (see [39]). One of papers mentioned in that survey deserves a special comment. We mean here the considerable progress towards the solution of problem (E_m) achieved by J. R. Isbell and Z. Semadeni in their joint paper [83]. As it is stated by the authors, inspiration came from B. Grünbaum (see [77]).

The main result of [83] reads as follows

Theorem 0.1.7. (Isbell, Semadeni). Let D be a complemented subspace of a Banach space X and suppose that D is isometrically isomorphic to the dual of a Banach space Y. Then there exists a minimal projection from X onto D. This is the case, in particular, if D is a reflexive subspace of X.

Compare this with the result of R. James' paper [85] (which appeared one year later, in 1964) concerning property (E): it is only in a reflexive space that each hyperplane (codimensiom one sobspace) enjoys property E.

Needless to say, Theorem (0.1.7) does not exhaust all the knowledge on problem (E_m) (see e.g. [22],[43]). Several results have been obtained on the existence of minimal projections onto subspaces which are not the duals of any space. For instance, no infinite dimensional subspace of c_o is a dual space (see [15],Corollary 2). Nevertheless, some of these subspaces admit minimal projection from c_o.

As regards the problem of uniqueness, survey [39] contains just a few theorems, which concern norm 1 projections.

This book is mainly devoted to the solution of uniqueness and strong unicity (see Chapter III) problems for minimal projections (with nonunit norm in general) in Banach spaces;it also exhibits the relevance of this problem to the uniqueness problem (U_{MP}) in mathematical programming and to the problem of characterization of Hilbert spaces.

For quite a long time already spaces with symmetric norm (for instance, l^n_p, l_p, L_p $(1 \le p \le \infty)$, c_o) have been used as a tester for new ideas in functional analysis,although the term "symmetric space" appeared relatively recently, in Singer's papers [175],[176]. (See e.g. Kadec-Pełczyński [91] or [123], or the papers on the interpolation of linear operators [173],[174], or papers on Banach lattices [1],[87];for detailed bibligraphy see [17],[25],[178]). Therefore, the problem of uniqueness of minimal projections deserves special interest in symmetric spaces. On the other hand, in symmetric space of finite dimension,the uniqueness problem for minimal projections is essentially a problem of mathematical programming.

Now, the question of uniqueness in problems of mathematical programming (we confine attention to linear programming)admits two approaches. The first of them consists in an analysis of the programming algorithm and requires its execution (see e.g. [59]). The other possibility is to inspect carefully the condition of the problem without executing the algorithm. This second way of approach has proved especially fruitful in the solution of several problems in matrix games (see e.g. [15],[23],[26], [69],[70],[129],[158]).

The interrelation between functional analysis and mathematical programming is well known. It suffices to resort for instance to the works of L. V. Kantorovich and his disciples.

In most of these papers, however, methods of functional analysis are applied in solving one or another problem of mathematical programming, whereas in the present book the direction is reversed. Methods of mathematical programming (in particular, the simplex method) are used in Chapter II to the solution of a functional analysis problem, namely, the problem of uniqueness of a minimal projections in a symmetric space of finite dimension.

In view of the natural relation between the two uniqueness problems (U_{MP}) and (U_{M}), established in this paper, it seems reasonable to hope that it might find an application in mathematical programming.

Let us now turn attention to the question of a characterization of Hilbert spaces within the class of all Banach spaces. This questions belongs among problems of functional analysis which have gained big popularity (see e.g. [58],[100],[161]). The following two questions have been posed in this connection by S. Banach [7] (The second question had been inspired by S. Mazur (see. [8], p.211).

(1.B) Let B be a Banach space ($\dim B \geq 3$) and $k \geq 3$ an integer. Suppose that all k-dimensional subspaces of B are mutually isometric. Is B necessarily isometric to a Hilbert space?

(1.B-M) In a Banach space B, if $x, y \in B$ are any elements with unit norm then there exists a linear isometry A carrying B onto itself and such that $Ax = y$. (A space with this property is sometimes called an isotropic space or a space with a transitive norm (see [50]).) Supposing

B to be separable, must B be isometric to a Hilbert space?

As shown by S. Rolewicz [167], the assumption of separability is necessary for a positive answer to problem (1.B-M).

According to the results of further research [131], if the space is infinite-dimensional, the requirement that the space norm is transitive cannot be replaced by partial transitivity. More precisely, every separable space B is a complemented subspace of a separable space X whose unit sphere contains a dense subset T such that for any $x, y \in T$ there is a linear isometry A of X onto X with $Ax = y$.

As to the first question, it is still lacking a complete solution in the case when $4 \le \dim B < \infty$. The second question is unanswered for $\dim B = \infty$. For finite dimensional spaces this problem has been positively solved by H. Auerbach (see [6], [7]). For more information the reader is referred to [61],[62],[76],[161].

Along with the questions (1.B) and (1.B-M) we will concern with the following two problems:

($1.B^{\circ}$) Let B be a Banach space and k a positive integer. Suppose that all subspaces of codimension k in B are isometric. Is B necessarily isometric to a Hilbert space?

Evidently, the problem ($1.B^{\circ}$) does not differ from the problem (1.B) in the case of $\dim B < \infty$. (If $4 \le \dim B < \infty$ then $k \le \dim B - 1$.)

(1.Od) Let B be a reflexive Banach space and k a positive integer. Suppose that each subspace of codimension k in B is the image of minimal projection $P_{(k)}$ with norm $1+\alpha$, where $\alpha \ge 0$ is a constant depending on B alone. (Then B is called (α,k)-space). Is B necessarily isometric to a Hilbert space? (The case $\alpha=0$, $k=1$ is treated in [7], p.254.)

It has to be remarked that problem (1.Od) is closely related to the well known result of Kakutani (see e.g. [58]) which states that if in a Banach space B with $\dim B \ge 3$ every subspace is the image of a projection B with norm $\|P\| = 1$, then B is isometric to a Hilbert space.

Problems (1.B) and (1.B-M) indicate that the task of characterization of Hilbert spaces among all Banach spaces unavoidably leads to investigations of the isometry group of Banach spaces involved. Investigation of that type has been originated by S. Banach with regard to spaces l_p, L_p ($1 < p < \infty$, $p \ne 2$), C(X). Further research has spread to spaces of complex-valued functions and sequences (see [63],[64],[112], [163] to mention just a few) on the one hand and the spaces with a basis and a symmetric norm [167], arbitrary symmetric coordinate spaces [29] and also certain nonsymmetric coordinate spaces defined by W. Orlicz [105],[106],[179], on the other. We have not mentioned here the literature concerning isometries in subspaces of classical spaces (see e.g. [111]), results on the groups of isometries of Banach spaces [5], [167] and, in particular, groups containing reflections [191].

In this paper we take up the problem of characterization of Hilbert spaces, chiefly, within the class of uniformly smooth strictly normed spaces. We will be concerned with problems (1.B°) and (1.Od), as well as with certain properties, hitherto ignored, of the isometry subgroup connected with a fixed subspace of the space under consideration.

__________ *** __________

We now expose in more detail the results of the book.

In Chapter I we pursue the uniqueness problem (U_m) in an arbitrary Banach space.

The main result of section 1 states that, in a three-dimensional space, a minimal projection with nonunit norm onto a codimension one subspace always is unique. As regards spaces of higher dimension, an analogous statesment is not true, in general; this is a result of K. Kursten, which we also present (see also section 3 of Chapter II).

In section 2 we examine certain properties of minimal projections. In this context, specific features of projections with nonunit norm are readily seen. The difference between these operators and projections of unit norm is made plain by two circumstances. Firstly, the latter always do attain their norm; and secondly, the question of key significance for the uniqueness of norm one projections onto a subspace D is the possibility of a unique norm-preserving extension of functionals on D to the whole B (see, for instance, Corollary I.2.20; also [13], Theorem 5). In other words, what is of main importance for the uniqueness of a projection with unit norm, is the "smoothness" of points of D in the space B. For minimal projection with nonunit norm, smoothness of that type is is also important, but only in the case when $\dim B/D \geq 2$. If $\dim B/D = 1$ ($\dim B \geq 4$), the fundamental part for the problem of uniqueness of minimal projection with nonunit norm is played by the concept of strict convexity of a subspace (see Theorem I.2.3 below).

In the same section we present V. M. Kadec's result on the linearity of a Lipschitz projection projection with unit norm.

Section 3 is mainly devoted to an inspection of the two diagrams

$$\begin{array}{ccc} & B & \\ {\scriptstyle P_1}\swarrow & & \searrow{\scriptstyle P} \\ K_1 & \xrightarrow[P_3]{} & D \end{array} \qquad\qquad \begin{array}{ccc} & B & \\ {\scriptstyle P}\swarrow & & \searrow{\scriptstyle P_2} \\ D & \xrightarrow[P_4]{} & K_2 \end{array} \qquad\qquad (0.1.16)$$

with P, P_i $(i=1,2,3,4)$ denoting minimal projections, $\|P\| > 1$, $\|P_2\| > 1$, $\|P_3\| > 1$, $\|P_1\| = \|P_4\| = 1$. It is pointed out, in what way questions concerning the uniqueness of P are influenced by the uniqueness of the P_i's.

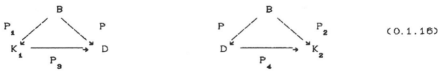

Also conditions are given under which projections P_3 and P_2 attain their norms.

The main object of discussion in this section are the uniqueness results for minimal projections in concrete spaces, whereas other facts of approximation theory (in particular, results of B. L. Chalmers, Z. Ciesielski [49], J. A. Mason [133]), as well as results on positive projections, are nearly not touched upon (here, as in the whole book, in fact).

In section 4 we consider the existence (and uniqueness) problem for a minimal projection onto a subspace D of a Banach space B provided there exists a (unique) minimal projection onto D from any subspace K containing D, with dim K/D = 1. It should be mentioned, in this connection, that in 1964 J. Lindenstrauss [120] constructed an example of a space B and a subspace D such that dim B/D = 2, $\rho(B,D) = 1$ and D admits no minimal projection from B, though it does admit a norm one projection for every subspace $K \supset D$ with dim K/D = 1.(See also [119], [122]).

The results of Chapter I make it apparent that, in solving problem (U_m) for subspaces of arbitrary codimension, the case of codimension one subspaces is of particular importance.

Chapter II exhibits the interplay between problems (U_m) for codimension one subspaces of finite-dimensional spaces with symmetric norm and a certain problem of linear programming in $\mathbb{R}^n$ (which we call the Bf-problem; see Definition II.2.1) with an objective function defined by a subspace. Problem (U_m) is then solved for l^n_∞, l^n_1, c_0 (and partially for l_1).

In section 1 of this chapter we give some complementary facts about symmetric spaces and we introduce terminology used in this paper in Chapter II alone.

In section 2 we associate with problem (U_m) a problem of mathematical programming defined by the space B and a functional $f \in B^* \setminus \{0^*\}$ (short Bf-problem) and we state conditions which ensure the admissibility of that construction. For the case of symmetric space we give a formula evaluating the norm of projection onto $D = f^{-1}(0)$.

In section 3 we solve the Bf-problem for the space $B = l^n_\infty$ $(n \geq 3)$ and subspace $D = f^{-1}(0)$, where $f \in B^* \setminus \{0^*\}$. As a consequence we get a solution of the problem of uniqueness of a minimal projection onto a codimension one subspace in l^n_∞.

In the section 4 we present a solution of the Bf-problem for l^n_1 $(n\geq 3)$ with use of the classical simplex method. We also give an example of solving the uniqueness problem for optimal strategy in a mathematical game by means of a certain Bf-problem.

In section 5 we solve the uniqueness problem of a minimal projection onto a codimension one subspace in l_1^n.

In section 6 we solve the uniqueness problem for minimal projections in c_o, and partially in l_1. To be precise, the problem in l_1 is solved for the case (so far unique) where one knows that a minimal projection onto a codimension one subspace does exists and has norm greater than one.

In section 7 we prove after [169] the following estimation

$$\Delta_1(l^p) \le \Delta_1(L^p[0,1]) \le 2^{|1/p-1|} \qquad (0.1.17)$$

for $1 < p < +\infty$, where

$$\Delta_1(X) = \sup(\rho(X,D): D \subset X, \text{ D is a closed hyperplane}) \qquad (0.1.18).$$

It is worth saying that the problem of calculating the constants $\rho(l^p,D)$ and $\rho(L^p[0,1],D)$ $(1<p<+\infty)$ has not been solved yet. (For more detailed information the reader is referred to [11] and [169]).

Section 8 is devoted to some results concerning projections onto subspaces of finite codimension. We prove the following estimations

$$\underline{\Delta}_k(L^p[0,1]) \le \Delta_1(L^p[0,1]) \text{ for } 1 \le p \le +\infty \qquad (0.1.19)$$

and

$$\bar{\Delta}_2(L^p[0,1]) > \Delta_1(L^p[0,1]) \text{ for } 1 \le p \le +\infty \qquad (0.1.20)$$

where

$$\bar{\Delta}_k(X) = \sup(\rho(X,D); D \subset X, \text{ codim } D = k) \ (k=1,2,\dots) \qquad (0.1.21)$$

and

$$\underline{\Delta}_k(X) = \inf(\rho(X,D): D \subset X, \text{ codim } D = k) \ (k=1,2,\dots). \qquad (0.1.22)$$

Next we dealt with the inequality

$$\bar{\Delta}_k(X) \le \bar{\Delta}_k(Y) \qquad (0.1.23)$$

for X being an Orlicz space of functions defined on the interval $[0,+\infty)$ and Y being a corresponding sequence Orlicz space.

In Chapter III we present various Kolmogorov type characterizations of minimal projections. These criteria do not concern only the problem of calculalating of minimal projections but also the problem of finding SUBA projections. Recall that a projection P_o of B onto D is called a strongly unique element of best approximation (briefly SUBA) with a constant $r>0$ if and only if

$$\|P\| \ge \|P_o\| + r \cdot \|P - P_o\| \qquad (0.1.24)$$

holds for every projection P of B onto D. It is clear that if P_o is a SUBA projection, then P_o is a unique minimal projection. The converse is not true (see Chapter III, section 3). We would like to stress that the results presented in this chapter are valid not only for the set of projections but for convex sets included in some space of compact operators. For this reason we do not restrict ourselves to the case of projections and we present the results in more general form.

In section 1 (preliminary section) we introduce terminology and notions which will be used only in this chapter.

In section 2 we work in the general case of the space $\mathcal{K}(B,D)$ of all compact operators going from a Banach space B into a Banach space D (not necessary a subspace of B). In particular we obtain a necessary condition for minimality of projections with norm greater than one which go onto finite dimensional subspaces (see Th.III.2.8).

In section 3 we discuss the problem of SUBA projections onto hyperplanes in l_1^n and l_∞^n. The results of this section may be treated as a developement of sections 5 and 6 of Chapter II. On the other hand they illustrate an application of theorems from section 2 in concrete cases.

In section 4 we concentrate on the space $\mathcal{K}(C(T,\mathbb{K}))$ of all compact operators going from the space $C(T,\mathbb{K})$ onto itself (As usual $C(T,\mathbb{K})$ denotes the space of $\mathbb{K}$- valued functions ($\mathbb{K} = \mathbb{R}$ or $\mathbb{K} = \mathbb{C}$) defined on a compact set T with the supremum norm.). In particular we characterize minimal projections with discrete carriers among all projections with discrete carriers.

Section 5 shows applications of the results of section 4 to the set $\mathcal{P}_D(C(T,\mathbb{K}),D,F)$ of all projections going from $C(T,\mathbb{K})$ onto finite dimensional subspace D with carriers contained in F.

Finally, section 6 deals with the case of generalized sequence spaces $c_o(T)$ and $l_1(T)$.

Chapter IV is devoted, on the one hand, to a study of isometries of a Banach space onto itself, and on the other, to characterization of Hilbert spaces within the class of uniformly smooth strictly normed Banach spaces, with aid of minimal projections.

In section 1 of Chapter IV it is shown that every isometry A of a Banach space B, having a nontrivial set B^A of fixed points such that $B = B_A \oplus B^A$ (where $B_A = (A-I)B$, I denoting the identity operator in B), induces a minimal projection onto B_A (see Theorem IV.1.1). We also consider some isometries generating minimal projections (see e.g. Example IV.1.8) and give conditions for B being the direct sum of B_A and B^A.

In section 2 we introduce a new functional ϕ_x in a uniformly convex strictly normed space. This functional is defined with use of minimal projections. In the case where its norm equals 1, it coincides with the Gateaux differential. A characterization of Hilbert spaces among uniformly smooth strictly normed spaces is obtained in terms of this functional.

In section 3 we prove that every isotropic space, i.e. space with a transitive norm, is an $(\alpha,1)$-space, with $\alpha \geq 0$. Moreover, we show that in a uniformly smooth isotropic space all subspaces of codimension 1 are mutually isometric.

In this connection, an examination of Rolewicz' isotropic space (see [167], IX, 6) shows that problem $(1.B^{o})$ has a negative solution for $k = 1$ whenever B is nonseparable. Besides, we get for $k = 1$ and $\alpha > 0$ a negative answer to the problem (1.0d).

The majority of results included in the present paper were published in 1975 - 85.

2. Terminology and notations

1. Let B be a Banach space over a field $\mathbb{K}$ ($\mathbb{K} = \mathbb{R}$ or $\mathbb{K} = \mathbb{C}$). We denote by $S_B(x;r)$ the sphere with centre $x \in B$ and radius $r > 0$, and by $W_B(x;r)$ the closed ball with centre x and radius r; thus

$$S_B(x;r) = \{y \in B: \|x - y\| = r\},$$

$$W_B(x;r) = \{y \in B: \|x - y\| \leq r\}.$$

By 0 we denote a zero element of B and we often write S_B and W_B instead of $S_B(0;1)$ and $W_B(0;1)$ (the unit sphere and ball in B).

As usual, B^* denote the dual of B (i.e. the space of all bounded linear functionals on B), and B^{**} is the second dual of B (i.e. $B^{**} = (B^*)^*$).

Let $\gamma: B \to B^{**}$ be the natural embedding, given by $(\gamma(x)f) = f(x)$ for $f \in B^*$ and $x \in B$. A Banach space B is reflexive if $\gamma(B) = B^{**}$.

A set of functionals $\mathfrak{X} \subset B^*$ is called total on a subspace $D \subset B$ if the equality $f(x) = 0$ holds for all $f \in \mathfrak{X}$ forces $x = 0$.

Let X and Y be subspace of B. If $B = X + Y$, $X \cap Y = \emptyset$, we write $B = X \oplus Y$ and call B the direct sum of X and Y. If , moreover , the norm of B satisfies the condition $\|x+y\|^p = \|x\|^p + \|y\|^p$ $(1 \leq p < \infty)$ (or $\|x+y\| = \max(\|x\|, \|y\|)$ for $x \in X$, $y \in Y$, we write $B = [X \oplus Y]_p$ (or $B = [X \oplus Y]_\infty$).

In a routine way we introduce the direct sum of k subspaces $X_1,\dots,X_k$ writing $B = \bigoplus_{i=1}^{k} X_i$.

Given any two subspaces X and Y in B, we denote by

$$(X\hat{,}Y) = \inf(\|x+y\|/\|x\|: x \in X\setminus\{0\}, y \in Y),$$

the inclination of X to Y.

2. Suppose that D is is a complemented subspace of B; that means, there exists a projection (idempotent continuous linear operator) from B onto Y. We denote by $\mathcal{P}(B,D)$ the set of all projections from B onto D and by $\rho(B,D)$ the relative projection constant, $\rho(B,D) = \inf\{\|P\|: P \in \mathcal{P}(B,D)\}$. For $P \in \mathcal{P}(B,D)$ we write crit $P = \{y \in S_B: \|Py\| = \|P\|\}$.

Let $\rho(B,D) = p$. The symbol $\Delta(B,D)$ denotes the subset of $\mathcal{P}(B,D)$ consisting of all minimal projections from B onto D, i.e. projections with norm equal to $\rho(B,D)$. If $\Delta(B,D) \neq \emptyset$, subspace D is called p-regular. (To indicate the value of $p = \rho(B,D)$, we sometimes write $\Delta^P(B,D)$ instead of $\Delta(B,D)$). If the set $\Delta(B,D)$ consists of a single element, the pair (B,D) is said to have the uniqueness property; in symbols card $\Delta(B,D) = 1$.

Let K be a complemented subspace of a Banach space B and let $P \in \mathcal{P}(B,K)$. Then E_P^K denotes the set $\{f \in S_{B^*}: \|f \circ P\| = \|P\|\}$. By B/K we denote the factor space modulo K and by $\operatorname{codim}_B K$ the dimension of B/K, i.e. the codimension of K in B.

3. Given a set $A \subset B$, we denote by $\bar{A}$ the closure of A (in the topology in B) and by Int A the interior of A (the set of all interior points of A); [A] is the span of A, i.e. intersection of all subspaces containing A.

A linear functional $f \in B^* \setminus \{0^*\}$ is said to be supporting for a set A (we also say: f is a support functional of A) if f attains its supremum over A at some point $x_o \in A$; i.e. $\sup f(A) = f(x_o)$. Whenever we speak of a support functional in B without specifying the set A, it is understood that "support" or "supporting" refers to W_B, the unit ball of B.

For any points $x_1, x_2 \in B$, we denote by $[x_1,x_2]$ the line segment with endpoints x_1,x_2, i.e. the set $\{x \in B: x = t\cdot x_1 + (1-t)\cdot x_2,\ 0 \le t \le 1\}$.

A Banach space is said to be strictly normed if its unit sphere contains no segments other that the trivial one (reducing to singletons).

If $D \subset B$ is a subspace of B and $\mathcal{K} \subset B$ is a set of functionals, we denote by $\mathcal{K}|_D$ the set of all functionals in $\mathcal{K}$ restricted to D.

The symbol $\mathcal{G}_x$ denotes the smoothness subspace at the point $x \in B\setminus\{0\}$ i.e., $\mathcal{G}_x = \{y \in B: g(x,y) = \lim_{t\to 0} t^{-1}\cdot(\|x+t\cdot y\| - \|x\|)$ exists$\}$.

Let $x \in X \setminus \{0\}$, X being a subspace of B. We call x a smoothness point of X if $\mathcal{G}_x \cap X = X$. If $\mathcal{G}_x \cap X = [x]$, the point x is called conical.

We say that the norm $\|\cdot\|$ is differentiable at a point $x \in B \setminus \{0\}$ in the direction of $y \in B$ if if $y \in \mathcal{G}_x$ (i.e. $g(x,y)$ exists). If the norm of

B is differentiable at every point $x \in S_B$, then X is called a smooth space.

A Banach space B is uniformly smooth if to any $\eta > 0$ there corresponds $\varepsilon(\eta) > 0$ such that, for any $x, y \in B$ ($\|x\|, \|x\| \geq \alpha, \alpha \geq 0$ is a const) with $\|x - y\| \leq \varepsilon(\eta)$, we have $\|x + y\| \cdot (1 + \eta) \geq \|x\| + \|y\|$. It is a known fact [39], that a uniformly smooth space is reflexive and smooth.

An element $x \in B \setminus \{0\}$ is said to be orthogonal to an element $y \in B$ (in symbols $y \perp x$)when $\|x\| = \inf \{\|x + \lambda \cdot y\| : \lambda \in \mathbb{R}\}$. Let $x \in B \setminus \{0\}$ and let D be a subspace of B. We write $x \perp D$ iff $x \perp y$ for all $y \in D$.

Given a convex set $K \subset B$, we denote by ext K the set of all extremal points of K, i.e. such that $x = (1/2) \cdot (x_1 + x_2)$ holds for no points $x_1, x_2 \in K$ other than x.

Let $\mathscr{X} \subset B^*$. We denote by $\mathscr{X}^\circ$ the set $\{x \in B: f(x) = 0 \text{ for all } f \in \mathscr{X}\}$.

For any $f \in B^*$, crit $f = \{z \in S_B : |f(z)| = \|f\|\}$.

C(K) is the space of all continuous real-valued functions on a compact metric space K, with the usual sup-norm. If $f \in C(K)$ is such that $\|f\| = |f(q)|$ for exactly one point q in K, then f is called a peak function.

4. $\mathbb{N}$ is the set of all positive integers. The symbol $\Gamma(r)$ stands for either $\Gamma(n) = \{1, 2, \ldots, n\}$ (when $r = n \in \mathbb{N}$) or $\Gamma(\infty) = \mathbb{N}$).

A sequence $\{x_i\}_{i \in \Gamma(r)}$ in a subspace $D \subset B$ is called a Schauder basis if each element $z \in D$ is uniquely represented as a sum $z = \sum_{i \in \Gamma(r)} \alpha_i \cdot x_i$, where $\alpha_i \in \mathbb{R}$ for $i \in \Gamma(r)$.

5. E^n denotes the Euclidean space of dimension n. Let $K \subset E^n$ be a convex body symmetric with respect to the origin. We can introduce a new norm in the n-space under consideration by means of the Minkowski metric with respect to ∂K, the boundary of K, thus obtaining an n-dimensional Banach space, which we denote by $B_{\partial K}$.

Points (and sets of points) of E^n will be marked by a '-sign: x',K' and so on; and we write simply x,K,... when regarding this objects as elements of $B_{\partial K}$.

Of course, every n-dimensional Banach space B arises from E^n by renorming with respect to the set S_B. Therefore we sometimes (mainly in Chapter II) employ notation $(\mathbb{R}^n, \|\cdot\|_B)$ for the space B.

6. $C_o(2\pi)$ (shortly C_o) denotes the space of all 2π-periodic real-valued functions with the sup-norm; π_n the subspace of C_o consisting of all trigonometric polynomials of degree $\leq n$ ($n \geq 1$). $D_n(t)$ is the n-th Dirichlet kernel,

$$D_n(t) = 1 + \sum_{k=1}^{n} \cos(k\cdot t) = \sin((n+(1/2)\cdot t))/\sin((1/2)\cdot t).$$

The symbol F_n will denote the Fourier projection (see (0.1.10)).

7. We use standard notation for classical spaces of functional analysis: l_p, l_p^n $(n\geq 1)$, $L_p(\Omega,\Sigma,\mu)(1\leq p\leq\infty)$, c_o, c, $C = C[0,1]$, (see e.g. [80], [80], [171]).

More specific terms, which appear only locally (in a single proposition, theorem, lemma) are defined in footness accompanying the proposition (theorem e.t.c.) in question.

8. All items (definitions, theorems, propositions, corollaries, le-lemmas, examples,remarks) are numbered consequtively within any section, without repetitions. And thus e.g. Definition II.2.1 opens the second section of Chapter II and is followed by Proposition II.2.2, Corollary II.2.3, Remark II.2.4, Theorem II.2.5 and so on. When reference is made to an item within the same chapter, the chapter number is omitted (e.g. we write: see Proposition 2.2). Seperate numbering is employed for formulas.

The symbol *** marks the end of each proof.

Chapter I

Problem of uniqueness of minimal projections in Banach spaces

§ 1. Uniqueness of a minimal projection with nonunit norm in a three dimensional Banach space

This section (and to some extent also section 4 of this chapter) is visibly distinguished among all sections of this book by its geometrical character. It contains a pretty large number of figures; in fact, there are more of then here than in all remaining section altogether. The reason for this is the authors' desire (reflecting also their taste) to make the exposition of the material as transparent as possible. At several points the proofs are not elaborated up to the very end, to avoid the danger of losing the essential idea in mess of oppressive details. It goes without saying that this is done only in case where the missing verification is of purely technical nature and does not exceed the abilities of a graduate student.

Under otherwise stated, through this section B denotes a three-dimensional Banach space and D is a two-dimensional subspace of B. Let P be a minimal projection of B onto D (in symbols , $P \in \Delta(B,D)$). We denote by P(crit P) the image under P of the set crit P = $\{y \in S_B : \|Py\| = \|P\|\}$.

Lemma I.1.1. Let $P \in \Delta(B,D)$. Then the set P(crit P) contains at least six points.

Proof. Obviously, crit P $\neq \emptyset$, by the compactness of the unit ball W_B. The assertions of the lemma are clearly true when $\|P\| = 1$. Thus assume that $\|P\| = p > 1$.

a) Suppose there are only 2 points, x and -x, in P(crit P). Take a point $z \in P^{-1}(x) \cap S_B$. By our assumption, the P-image of S_B touches the set $S_D(0;p)$ at two points only, x and -x (see Fig. 1 and 2).

Let $P_t : B \to D$ be the projection anihilating the straight line parallel to the segment with endpoints z and $(1-t)\cdot x$, where t is a sufficiently small positive number. Taking into account that the image of W_B under projection onto D is a bounded convex plane set, having 0 as the centre of symmetry, and applying the standard compactness argument, it is not difficult to check that there is $t>0$ such that $P_t(S_B)$ does not touch $S_D(0;p)$ at all. Consequently $\|P_t\| < p$, contrary to the minimality of $\|P\|$; we have arrived a contradiction.

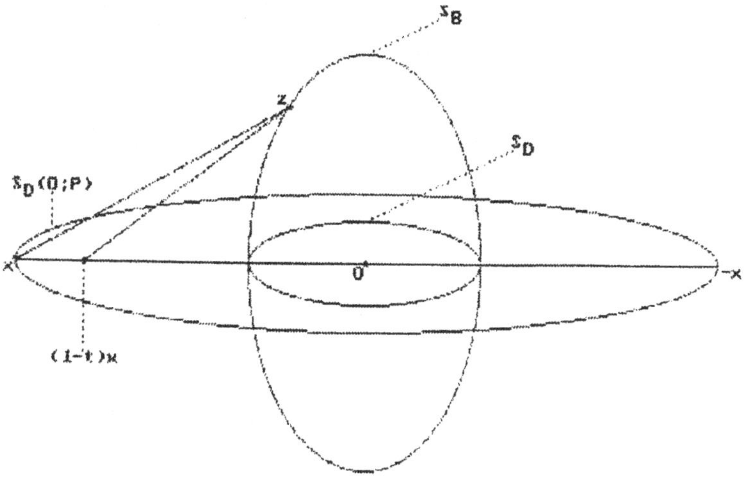

Figure 1

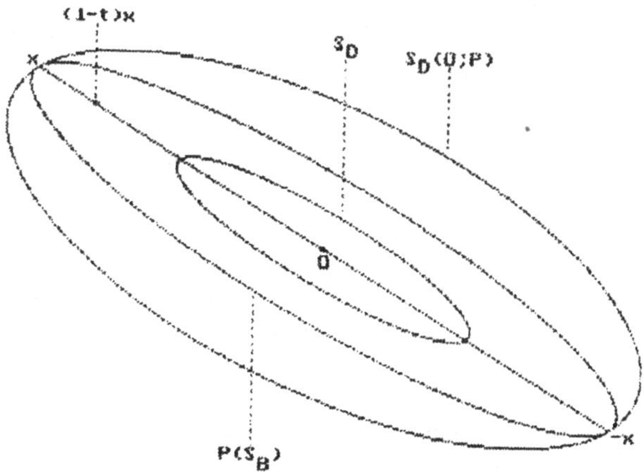

Figure 2

b) Now suppose that P(crit P) contains of 4 points only, $x, y, -x, -y$. Pick points z_1, z_2 in S_B, $z_1 \in P^{-1}(x)$, $z_2 \in P^{-1}(y)$. Out of the four points z_1, z_2, $-z_1$, $-z_2$, either z_1 and z_2 or z_1 and $-z_2$ lie on the same side of D.

There is no loss of generality in assuming that z_1 and z_2 are on one side. Let $P_t: B \to D$ be the projection annihilating the line parallel to the segment with endpoints z_1 and $x - t\cdot(x+y)$, where t is a sufficiently small positive number. (See Fig. 3).

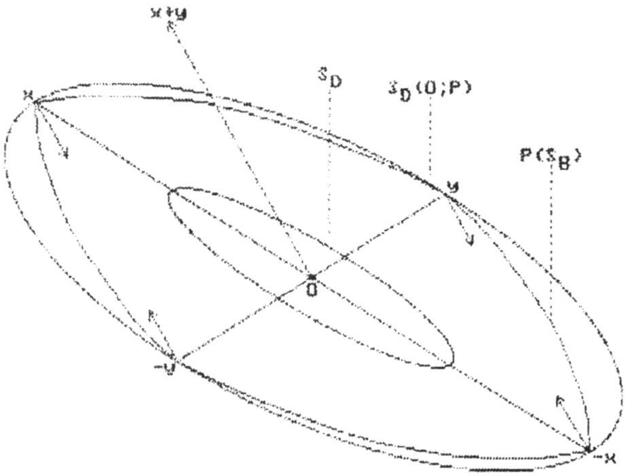

Figure 3

Again it is not hard to see that there exists $t > 0$ such that $P_t(S_B)$ lies in the interior of $W_D(0;p)$. Hence $\|P_t\| < p$, contrary to the minimality of $\|P\|$. The resulting contradiction prove the statement.***

The next lemma is proved by fully elementary reasoning.

Lemma I.1.2. Let u be a boundary of a centrally symmetric bounded plane figure $W' \subset E^2$. Let x'_1, x'_2, x'_3 be three distinct points of u with the following properties:

a) x'_1, x'_2, x'_3 lie on one side of a certain diameter (i.e. a line segment passing through the symmetry centre of W' with endpoints on u) of W';

b) x'_2 is an interior point of the shortest arc in u that connects the points x'_1 and x'_3;

c) there do not exist parallel (or coincide) straight line segment contained in u and containing the points x'_1 and x'_3.

Suppose that there is fixed a direction in the plane of W' and that three points are arbitrarily shifted along this direction, x'_1 and x'_3 in the same sense and x'_2 in the opposite. Then at least one of these points is moved off W'. (See Fig.4 and 5).

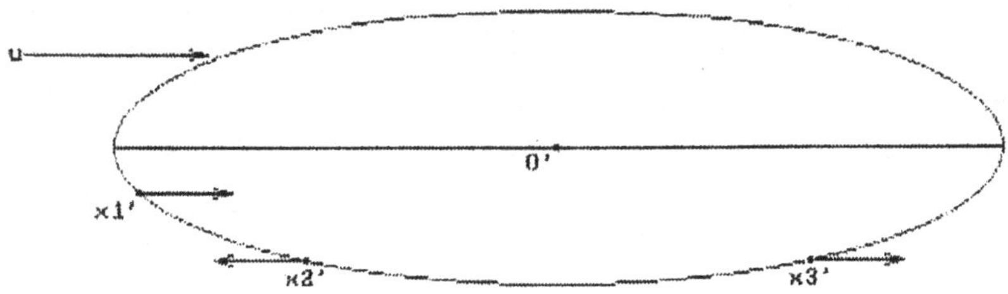

Figure 4

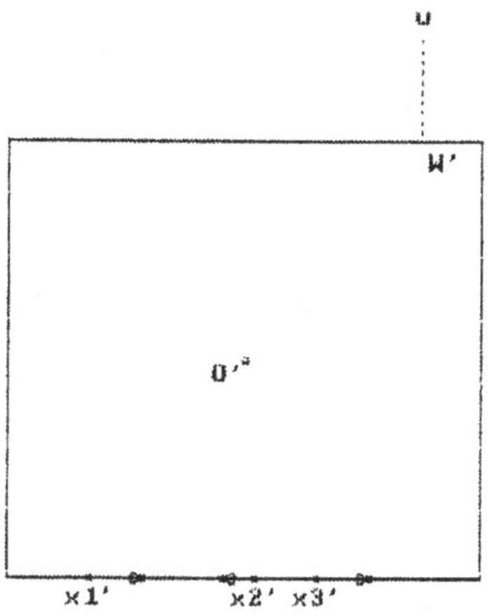

Figure 5

Theorem I.1.3. Let D be a two-dimensional subspace of a three dimensional Banach space B. If $\rho(B,D) = p > 1$, then the minimal projection of B onto D is unique; i.e. card $\Delta(B,D) = 1$.

Proof. Let $P \in \Delta(B,D)$. According to Lemma 1.1, the set P(crit P) consists of at least 6 points. The space B is partionned by D into two half spaces.

Every point in P(crit P) will be labelled with one of the superscripts (+) and (-) according as it is the image of a point (points) of one half-space or the other. Since $\|P\| > 1$, each point in P(crit P) can be assigned only one index.

Assume that there exists a diameter d of the ball $W_D(0;p)$ (To be rigorous, we are talking about the ball $W_D(0;p) \subset E^2$, here and in the sequel.) such that (+)-points lie on one side of it. Arguing as in the proof of Lemma 1.1, if we take x and y to be the endpoints of the shortest arc in the boundary of $W_D(0;p)$ containing all (+)-points, we can easily verify that projection P is not minimal - a contradiction (see Fig.6).

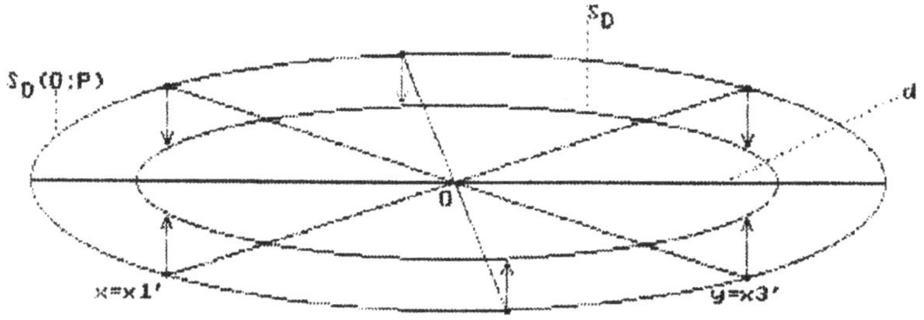

Figure 6

We hence conclude that there exists a diameter d of $W_D(0;p)$ and points $z_1^{(+)}$, $z_2^{(+)}$, $z_3^{(+)}$ $\in$ P(crit P) lying on one side of d such that $z_2^{(-)}$ belongs to the shortest arc in $S_D(0;p)$, the boundary of $W_D(0;p)$, which connects $z_1^{(+)}$and $z_3^{(+)}$. We claim that $z_1^{(+)}$ and $z_3^{(+)}$ cannot both belong to two parallel segments (let alone one line segment) contained in $S_D(0;p)$. Indeed, assume the contrary. Then clearly the point $-z_1^{(+)}$ has index (-) and there is a straight line portion of $S_D(0;p)$ with points $z_1^{(-)}$ and $z_3^{(+)}$ lying on it (and another one containing $z_1^{(+)}$and $z_3^{(-)}$). (see Fig.7).

Then the ball W_B contains the segment $[z_1,z_3]$, where z_1 and z_3 are inverse images of $z_1^{(+)}$ and $z_3^{(-)}$ under projection P. On the other hand, it is clear (see Fig.8) that there exists a point $x_0 \in [z_1,z_3] \cap [z_3^{(-)},z_1^{(+)}]$ Hence, $W_D(0;p) = W_D(0;1)$, which means that $\|P\| = 1$, contrary to the assumption of the theorem. This proves the claim, showing that the points $z_1^{(+)}$, $z_2^{(-)}$,$z_3^{(+)}$ satisfy the condition of Lemma 1.2.

Now, assuming that projection P is not unique of norm p, we soon arrive at the conclusion that there exists a direction in D such that small shifts of the three points ($z_1^{(+)}$and $z_3^{(+)}$ in one sense, $z_2^{(-)}$ in the other) do not carry these points outside $W_D(0;p)$. But it contradicts Lemma 1.2.***

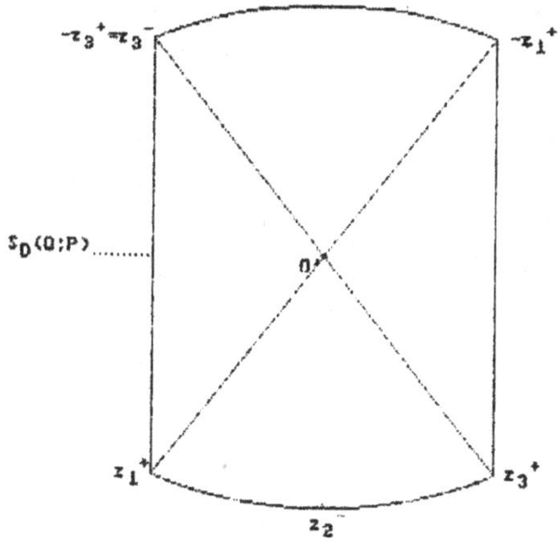

Figure 7

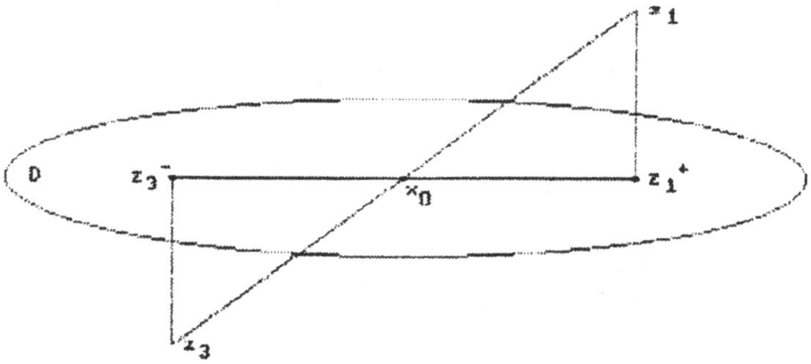

Figure 8

Remark I.1.4. In arbitrary Banach spaces (of dimension greater than 3, of course), a minimal projection onto a p-regular subspace of codimension 1 is not unique, in general, even if $p > 1$.

As an example one can consider the space l_∞^n, $n \geq 4$ (see below, Remark II.3.1).

An analogous result has been obtained independently by K. Kürsten (see [149], Proposition 2). More precisely, he has proved the following

Proposition I.1.5. **(Kürsten).** Let D be a p-regular subspace of B, $p > 1$. Let $X = [B \oplus \mathbb{R}]_\infty$, $K = [D \oplus \mathbb{R}]_\infty$. Then K is p-regular subspace of X and card $\Delta(X,K) = +\infty$.

Proof. Let $P \in \Delta^{P}(B,D)$. Take a functional $f \in B^{*}$ with $f^{-1}(0) \supset P^{-1}(0)$. Let $\varepsilon_{o} = (\|P\| - 1)\cdot(\|f\|\cdot\|I-P\|)^{-1}$. For each ε between 0 and ε_{o} define an operator $P_{\varepsilon}: X \to K$ by

$$P_{\varepsilon}(x,\lambda) = (Px, \lambda+\varepsilon\cdot f((I-P)x)) \text{ for } x \in B,\ \lambda \in \mathbb{R}. \qquad (1.1.1)$$

Observe that $P_{\varepsilon}(x,\lambda) = (x,\lambda)$ whenever $x \in D$, $\lambda \in \mathbb{R}$. It is also clear that P_{ε} is an idempotent linear operator. Since

$$\|P_{\varepsilon}\| = \max\{(\|P\|, |\lambda + \varepsilon\cdot f((I-P)x)|): z \in S_{B}\},$$

P_{ε} is a projection onto K and, moreover, $P_{\varepsilon} \in \Delta(X,K)$.
To see this, assume that there exists a projection $P_{*}: X \to K$, $\|P_{*}\| < p$. Let P_{o} be the projection from $K = [D \oplus \mathbb{R}]$ onto D given by $P_{o}(x,\lambda) = x$ for $x \in D$, $\lambda \in \mathbb{R}$. Then $\|P_{o}\| = 1$ and the operator $P_{o} \circ P_{*}$ is a projection of X onto D with norm $\|P_{o} \circ P_{*}\| < p$. Consider $P_{o} \circ P_{*}|_{B}$ (operator $P_{o} \circ P_{*}$ restricted to B); it is a projection from B onto D with norm less than p. Consequently D is not p-regular, and this contradicts to the hypothesis of the proposition. Clearly enough, card $\Delta(X,K) = +\infty$. ***

§ 2. Certain properties of minimal projections

Let D be a complemented subspace of a Banach space B and let $P \in \mathcal{P}(B,D)$. The main objective of this section is to exhibit conditions which provide the existence and uniqueness of a minmal projections onto D. It turns out that the key role is played by sets which can be viewed as analogues to of the set P(crit P) dealt with in section 1. We denote this sets by M_{P}^{D} and $\bar{M}_{P}^{D}$:

$$M_{P}^{D} = P(S_{B}) \cap S_{D}(0; \|P\|) ;$$

$$\bar{M}_{P}^{D} = \overline{P(S_{B})} \cap S_{D}(0; \|P\|) . \qquad (1.2.1)$$

When dealing with the uniqueness problem (U_{m}) in the case of dim $B \geq 4$ we soon discover the significance of the smoothness sets of unit spheres in D and B. Denote by sm S_{D} and sm S_{B},

$$\text{sm } S_{D} = \{y \in S_{D}: \mathcal{J}_{y} \cap D = D\};$$

$$\text{sm } S_{B} = \{y \in S_{B}: \mathcal{J}_{y} \cap B = B\}. \qquad (1.2.2)$$

Throughout this section we write E_{P} in place of E_{P}^{B}. Recall that

$$E_{P} = \{f \in S_{B^{*}}: \|f \circ P\| = \|P\|\}. \qquad (1.2.3)$$

Let us also recall that, given a point $x \in B \setminus \{0\}$, we denote by $A^{B}(x)$ the set $\{f \in S_{B^{*}}: f(x) = \|x\|\}$.

Lemma I.2.1. Let $D \subset B$ and let $P \in \mathcal{P}(B,D)$. Then

$$E_P \supseteq \bigcup_{y \in \bar{M}_P^D} A^B(y) \quad \text{if } \bar{M}_P^D \neq \emptyset\ ; \tag{1.2.4}$$

$$\text{\&} \quad E_P = \bigcup_{y \in \bar{M}_P^D} A^B(y) \quad \text{if } \dim D < +\infty. \tag{1.2.5}$$

Proof. Let $\bar{M}_P^D \neq \emptyset$. Pick $y \in \bar{M}_P^D$. By definition of $\bar{M}_P^D$, there exists a sequence $\{z_n\} \subset S_B$ such that $P(z_n) \to y$ as $n \to \infty$. Let $f \in A^B(y)$. Then $\lim_{n\to\infty}|f(P(z_n))| = |f(y)| = \|y\| = \|P\|$. On the other hand, $\|f \circ P\| \le \|f\| \cdot \|P\| = \|P\|$, i.e. $\|f \circ P\| = \|P\|$ and hence $f \in E_P$.

Now assume that $\dim D < \infty$. If $\|P\| = 1$ then $\bar{M}_P^D = S_D$ and (1.2.5) is obvious. Thus let $\|P\| > 1$. Take any functional $f \in E_P$. We can find a sequence $\{z_n\} \subset S_B$ such that $(f \circ P)(z_n) \to \|P\|$ as $n \to \infty$. Since D is finite-dimensional, there is a subsequence z_{n_k} with $\{P(z_{n_k})\}$ converging to a certain point y. Clearly, $y \in \overline{P(S_B)}$. Then $\lim_{n\to\infty} (f \circ P)(z_{n_k}) = f(y) = \|P\|$. Since $\|f\| =$ and $\|y\| \le \|P\|$, we get $\|y\| = \|P\|$ and thus $f \in A^B(y)$.***

Remark I.2.2. If D is not a reflexive subspace then $E_P \neq \bigcup_{y \in \bar{M}_P^D} A^B(y)$, even in the case $\|P\| = 1$.

Indeed, let $\|P\| = 1$. Then $\bar{M}_P^D = S_D$. Choose a functional $f \in S_{D^*}$, which does not attain its norm on D and extend it to a functional $\bar{f}$ on B with the same norm. Note that $\|\bar{f} \circ P\| = 1$(for $\bar{f} \circ P$ is an extention of f onto B). Hence, $\bar{f} \in E_P$. However, there is no $y \in S_D$ such that $\bar{f} \in A^B(y)$, otherwise the functional $f = \bar{f}|_D$ would attain its norm at y.***

Lemma 2.1 directly implies

Corollary I.2.3. Let $P \in \mathcal{P}(B,D)$. If crit $P \neq \emptyset$ then

$$E_P \supseteq \bigcup_{y \in P(\text{crit } P)} A^B(y). \tag{1.2.6}$$

Lemma I.2.4. Let B be a reflexive space, let $D \subset B$ and $P \in \mathcal{P}(B,D)$. Suppose that $\bar{M}_P^D \neq \emptyset$. Then

(i) crit $P \neq \emptyset$;

(ii) $E_P = \bigcup_{y \in P(\text{crit } P)} A^B(y)$.

Proof.(i) Take $y \in \bar{M}_P^D$ and $f \in A^B(y)$. By the definition of E_P, $f \in E_P$. Since B is reflexive, the functional $f \circ P$ attains its norm on B i.e. $(f \circ P)(z_1) = \|P\|$ for a certain $z_1 \in S_B$. But then $\|Pz_1\| = \|P\|$.

(ii) Choose $f \in E_P$. As in (i), we find $z_1 \in S_B$ with $f(P(z_1)) = \|P\|$ and $\|Pz_1\| = \|P\|$. Write $y = Pz_1$. Then clearly $f \in A^B(y)$. This gives an inclusion in (ii); the other inclusion is ensured by Corollary 2.3.***

Corollary I.2.5. Let D be a subspace of B and suppose that the set of minimal projections $\Delta(B,D)$ is nonempty. Suppose also that crit $P \neq \emptyset$ for each $P \in \Delta(B,D)$. Then for every $k \in \mathbb{N}$ and every $P_1, \quad ,P_k \in \Delta(B,D)$ there exists a vector $z \in S_B$ at which $P_1,\dots,P_k$ attain their norms.

Proof. Consider the operator $P_{k+1} = (1/k)\cdot\sum_{i=1}^{k} P_i$. Obviously, P_{k+1} is a projection and $\|P_{k+1}\| \le \rho(B,D)$. Consequently $P_{k+1} \in \Delta(B,D)$. By hypothesis, there exists $z_o \in$ crit P_{k+1}. It is not hard to check that $z_o \in$ crit P_i for $i=1,\dots,k$.***

Remark I.2.6. Let D be a finite-dimensional subspace of B. According to Theorem 0.1.2 and Lemma 2.4, the conditions of Corollary 2.5 are certainly satisfied whenever B is reflexive.

The next lemma shows that a minimal projection need not attain its norm, even when projecting onto a finite-dimensional subspace.

Lemma I.2.7. The Fourier projection F_n from $C_o(2\pi)$ onto π_n does not attain its norm on $C_o(2\pi)$.

Proof. Assume it does. Then there is an element $x_o \in S_B$, where $B = C_o(2\pi)$, such that $\|F_n\| = \|F_n(x_o)\|$. Since $F_n(x_o)$ is a continuous function, $\|F_n\| =$ $= |(F_n(x_o))(y_o)|$ at a point $y_o \in [0;2\pi]$.

Formulas (0.1.10) and (0.1.15) yield

$$\left|(1/2\pi)\cdot\int_0^{2\pi} x_o(t+y_o)\cdot D_n(t)\,dt\right| = (1/2\pi)\cdot\int_0^{2\pi} |D_n(t)|\,dt,$$

and consequently $\int_0^{2\pi} (1 - |x_o(t+y_o)|)\cdot|D_n(t)|\,dt = 0.$

Since $|D_n(t)| > 0$ almost everywhere on $[0;2\pi]$, we get $|x_o(t+y_o)| = 1$ for almost all $t \in [0;2\pi]$ and so, by continuity, either $x_o(t) = 1$ or $x_o(t)=-1$. In either case $x_o \in \pi_n$. Thus $F_n(x_o) = x_o$ and $\|F_n\| = 1$, contrary to Theorem 0.1.2.***

Theorem I.2.8. Let D be a p-regular ($p > 1$) uniformly convex subspace of a Banach space B. (Recall that a subspace D is called uniformly convex if the conditions $\|x_n\| = \|y_n\| = 1$, $\lim_{n\to\infty} \|x_n + y_n\| = 2$ yield $\lim_{n\to\infty} \|x_n - y_n\| = 0$.)

Then

$$(P_1^{-1}(0),\hat{}\,P_2^{-1}(0)) = 0.\text{(see Intr., section 2)} \tag{1.2.7}$$

for any two projections P_1, $P_2 \in \Delta(B,D)$.

Proof. Assume, contrary to the assertion, that there exist $P_1, P_2 \in \Delta(B,D)$ with $(P_1^{-1}(0),\hat{}\,P_2^{-1}(0)) = \beta \neq 0$ and consider the projection $P_3 = (1/2)\cdot(P_1 + P_2) \in \Delta(B,D)$. We can find a sequence $(z_n) \subset S_B$ such that $\lim_{n\to\infty} \|Pz_n\| = p$. Since

$$0 \le p - \|P_1 z_n\| + p - \|P_2 z_n\| \le |2\cdot p - \|P_1(z_n) + P_2(z_n)\|| = 2p - \|2\cdot P_3(z_n)\|,$$

we get

$$\lim_{n\to\infty} \|P_i z_n\| = p, \ (i=1,2). \tag{1.2.8}$$

Let $\|P_i z_n\| = p - \varepsilon_n^i$, $\varepsilon_n^i \ge 0$ $(i=1,2;n=1,2,\ldots)$.

Write $z_n^i = P_i z_n/\|P_i z_n\|$. Then $P_i z_n = z_n^i\cdot p - \varepsilon_n^i\cdot z_n^i$ $(i=1,2;n=1,2,\ldots)$.

Now, we have $\|z_n^i\| = 1$, so that

$$2 \ge \|z_n^1 + z_n^2\| = \|(1/p)\cdot(P_1 z_n + P_2 z_n + \varepsilon_n^1\cdot z_n^1 + \varepsilon_n^2\cdot z_n^i)\| \ge$$

$$\ge \ |\|(2/p)\cdot P_3 z_n\| - (1/p)\cdot\|\varepsilon_n^1\cdot z_n^1 + \varepsilon_n^2\cdot z_n^2\| \quad \text{and}$$

$$\lim_{n\to\infty} 2\cdot\|P_3 z_n\| = 2p,\ \lim_{n\to\infty} \|\varepsilon_n^1\cdot z_n^1 + \varepsilon_n^2\cdot z_n^2\| = 0; \text{hence}$$

$$\lim_{n\to\infty} \|z_n^1 + z_n^2\| = 2.$$

By the uniform convexity of D, $\lim_{n\to\infty} \|z_n^1 - z_n^2\| = 0$.

Since $\|P_1 z_n - P_2 z_n\| = \|p\cdot z_n^1 - \varepsilon_n^1\cdot z_n^1 - p\cdot z_n^2 + \varepsilon_n^2\| \le p\cdot\|z_n^1 - z_n^2\| + \varepsilon_n^1 + \varepsilon_n^2$, we see that

$$\lim_{n\to\infty} \|P_1 z_n - P_2 z_n\| = 0. \tag{1.2.9}$$

On the other hand, for any element $z \in S_B$ we have

$$P_i z = z + t_z^i\cdot y_z^i, \tag{1.2.10}$$

where $t_z^i \in \mathbb{R}$, $y_z^i \in P_i^{-1}(0) \cap S_B$ $(i=1,2)$. Thus (whenever $t_z^i \neq 0$),

$$\|P_1 z - P_2 z\| = \|t_z^1\cdot y_z^1 - t_z^2\cdot y_z^2\| = |t_z^1|\cdot\|(t_z^1\cdot y_z^1/t_z^1) + (-t_z^2\cdot y_z^2/t_z^1)\| \ge$$

$$|t_z^1|\cdot(P_1^{-1}(0),\hat{}\,P_2^{-1}(0)).$$

Let $p = 1 + t_o$, $t_o > 0$, and let $\hat{\varepsilon} = t_o/2$. In view of (1.2.8) there exists $n_o \in \mathbb{N}$ such that $\varepsilon_n^i < \hat{\varepsilon}$ for all $n > n_o$ $(i=1,2)$.

Consequently, the inequality $\|P_i z_n\| = 1 + t_o - \varepsilon_n^i > 1 + (t_o/2)$ holds for $n > n_o$. Hence, by (1.2.10),

$$|tz_n^i| > (1/2)\cdot t_o \quad (i=1,2).$$

Eventually, we get for $n > n_o$

$\|P_1 z_n - P_2 z_n\| > (1/2)\cdot t_o \cdot \beta > 0,$

which contradicts (1.2.9) and concludes the proof of the theorem. ***

Remark I.2.9. Relation (1.2.7) does not force the existence of an element $x \neq 0$ annihilating two projections i.e. such that $x \in P_1^{-1}(0) \cap P_2^{-1}(0)$.

Sometimes, however, it is true (see Corollary 2 12 below).

Proposition I.2.10. Let D be a p-regular ($p \geq 1$) subspace of B. Suppose that

(i) crit $P \neq \emptyset$ for every $P \in \Delta(B,D)$;

(ii) $Pz \in \text{ext}\, W_B(0;p)$ for every $P \in \Delta(B,D)$ and $z \in$ crit P.

Then, given any two projections in $\Delta(B,D)$, there exists a point $x \in S_B$ at which these two projections coincide.

Proof. For p = 1 the statement is obvious. Thus let $p > 1$. Choose any two projections $P_1, P_2 \in \Delta(B,D)$. Then $P_3 = (1/2)\cdot(P_1 + P_2) \in \Delta(B,D)$. By Corollary 2.5 there is an element $z_o \in S_B$ at which P_1, P_2, P_3 attain their norms. Hence, $P_1 z_o = P_2 z_o$; otherwise $P_3 z_o$ would not be an extremal point of the ball $W_D(0;p)$, contrary to the assumption (ii). ***

Corollary I.2.11. Let D be a complemented, strictly normed, reflexive subspace of B. Suppose that B/D is also reflexive. Then

(a) $\Delta(B,D) \neq \emptyset$;

(b) for any P_1, $P_2 \in \Delta(B,D)$ there exists $x \in S_B$ at which P_1, P_2 coincide, provided that $\bar{M}^D_{P_1} \neq \emptyset$ and $\bar{M}^D_{P_2} \neq \emptyset$.

Proof. By Theorem 0.1.7, D is a p-regular subspace of B (for a certain $p \geq 1$). Since the subspace D and the factor-space B/D are both reflexive, the space B is also reflexive (see [60], Ch.II, Exercise 20). In view of Lemma 2.4 and the assumption that D is strictly normed, conditions of Proposition 2.10 are satisfied, and this implies the statement b). ***

Corollary I.2.12. Let D be a finite-dimensional, strictly-normed subspace of a reflexive space B. Then $\Delta(B,D) \neq \emptyset$ and for any $P_1, P_2 \in \Delta(B,D)$ there is a point $x \in S_B$ at which P_1 and P_2 coincide.

Proof. It suffices to observe that, D being of finite dimension, the set $\bar{M}^D_P$ is nonempty, for any $P \in \Delta(B,D)$. ***

Theorem I.2.13. Let D be a uniformly convex subspace of B, with dim B/D=1, and suppose that $\rho(B,D) = p > 1$. Then

$$\text{card}\, \Delta(B,D) = 1.$$

Proof. A uniformly convex space always is reflexive and strictly normed (see e.g. [58]). Therefore, in view of Corollary 2.11, $\Delta(B,D) \neq \emptyset$.

Take any two projections P_1, $P_2 \in \Delta(B,D)$. Since dim B/D = 1, the subspaces $P_1^{-1}(0)$ and $P_2^{-1}(0)$ are one dime sional. Thus, by Theorem 2.8, $P_1^{-1}(0) = P_2^{-1}(0)$, which means that $P_1=P_2$. (See also Theorem 12 of [78]). ***

Corollary I.2.14. Let $D \subset B$ be a uniformly convex subspace with dim B/D = 1. Suppose that

(*) $S_D \subset \text{sm } S_B$.

Then card $\rho(B,D) = 1$.

Proof. Let $\rho(B,D) = p$. If $p > 1$ then the claim follows from Theorem 2.13. If $p = 1$, the statement follows, for instance, from Theorem 1 of [144]. One might also apply Corollary 2.20. (see below taking $\mathscr{K}^D = D^*$). ***

Remark I.2.15. a) Conditions for uniform convexity in concrete spaces are given e.g. in [81],[82],[92],[124].

b) Instead of imposing condition (*) for the uniqueness of minimal projection, in the case $p = 1$, one can use various specific versions of Corollary 2.20 (e.g. demand that B be a smooth space; see also [50]).

In the sequel we shall need the following result, due to Morris and Cheney (see [41], Th.5 or Th.III.2.8).

Theorem I.2.16. Let $P \in \Delta(B,D)$, dim $D < \infty$. Then either $\|P\| = 1$ or every subset $E \subset E_P$ (see (1.2.3)) such that

$$E \cap (-E) = \emptyset \text{ and } E_P = E \cup (-E) \tag{1.2.11}$$

is linearly dependent over D.

This theorem jointly with Lemmas 2.1 and 2.4 result in

Theorem I.2.17. Let $P \in \Delta(B,D)$, dim $D < \infty$. Let $\|P\| > 1$. Then

(i) the set $\bigcup_{y \in M_D^{-P}} A^B(y)|_D$ is total on D ;

(ii) if B is reflexive then $\bigcup_{y \in P(\text{crit } P)} A^B(y)|_D$ is total on D.

Example I.2.18. Let $D \subset B$, dim B = 3, and suppose that D is p-regular in B, $p > 1$. Let $P \in \Delta(B,D)$. By Theorem 2.17 the set $\bigcup_{y \in P(\text{crit } P)} A^B(y)|_D$. is total on D. Notice that dim D = 2 (because $p > 1$). Therefore taking

$$\mathscr{K}^1 = ((A^B(z_1^+)|_D) \cup (A^B(z_3^+)|_D)) \subset \bigcup_{y \in P(\text{crit } P)} (A^B(y)|_D),$$

where z_1^+ and z_3^- are the same points defined in the proof of Theorem 1.3, we can verify without dificulty that the set $\mathscr{K}^1$ is total on D.

Theorem I.2.19. Let D be a p-regular ($p \geq 1$) subspace of a Banach space B. The following conditions are necessary and sufficient for the equality card $\Delta(B,D) = 1$: there exists a set $\hbar^D \subset D^*$ with the properties

$1^{\circ\circ}$ $\hbar^D$ is total on D;

$2^{\circ\circ}$ for any $f \in \hbar^D$ and any $P_1, P_2 \in \Delta^P(B,D)$,

$$f\circ P_1 = f\circ P_2. \tag{1.2.12}$$

Proof. Necessity is obvious; it is enough, e.g., to take S_{D^*} as $\hbar^D$. For the proof of sufficiency, take any $P_1, P_2 \in \Delta(B,D)$ and suppose that $P_1 \neq P_2$. Then there exists $x \in P_1^{-1}(x)$ with $P_2x \neq 0$. By assumption $2^{\circ\circ}$ $(f\circ P_2)(x) = (f\circ P_1)(x)$ for all $f \in \hbar^D$. According to $1^{\circ\circ}$, $P_2x = 0$, in contradiction with the previous statement.***

Corollary I.2.20. Let D be a 1-regular subspace of B. Let $\hbar^D \subset D^*$. The following conditions are jointly sufficient for the equality card $\Delta(B,D) = 1$:

(i) $\hbar^D$ is total on D ;

(ii) every functional $f \in \hbar^D$ has the property (U) (uniqueness of norm preserving extention).

Proof. Just observe that, given any $f \in \hbar^D$ and $P_1, P_2 \in \Delta(B,D)$, the functionals $f\circ P_1$ and $f\circ P_2$ are extentions of f to all of B, preserving the norm; hence, by (ii), condition $2^{\circ\circ}$ of Theorem 2.19 is fulfilled.***

Remark I.2.21. Corollary 2.20 is a generalization of Theorem 1 of [144].

Example I.2.22. Let $B = l_p$ or $B = l_p^n$ $(n\geq 3, 1<p<\infty)$. Let D be a subspace of B, $\dim B/D = 1$. Then

a) there exists a unique minimal projection P of B onto D;

b) if $B = l_p$ and D is isometrically isomorphic to B then $\|P\| = 1$;

c) if $B = l_p$ and D is not isometrically isomorphic to B then $1\leq\|P\|\leq 2$.

Proof. a) First observe that a minimal projection onto D certainly exists, since the spaces l_p and l_p^n $(n\geq 3, 1<p<\infty)$ are uniformly convex (see e.g. [60], Ch. V, Sec.12) and hence reflexive. They are also smooth. Thus by Corollary 2.14, a minimal projection must be unique.

b) It has been shown by A. Pełczyński ([159], Th.2) that a subspace D, which is isometrically isomorphic to l_p, always admits a projection onto D of norm 1.

c) This follows from a result of Petunin and Levin [113] stating that the norm of a minimal projection onto a reflexive subspace of codimension n never exceedes n+1.***

Remark I.2.23. As it is seen from Lemma 2.7, the set M_P^D can be empty, even if P is a minimal projection and $\bar{M}_P^D \neq \emptyset$ (D being of finite dimension). On the other hand, one can find examples of situations when $\bar{M}_P^D = \emptyset$ and B is reflexive.

Problem I.2.24. Let $P \in \Delta(B,D)$. Does anyone of the following two conditions ensure that $\bar{M}_P^D \neq \emptyset$? The conditions are:

a) D is uniformly convex, B is reflexive;

b) D is strictly normed and reflexive, dim B/D = 1.

Proposition I.2.25. Let D be a superreflexive subspace of a space B, of codimension 1.(Recall that a space B is superreflexive (see e.g.[88],pp. 109-110) if for any nonreflexive space X and $\lambda \geq 1$ and any finite dimensional subspace $D_1 \subset X$ there is no subspace $D_2 \subset B$ (dim D_1 = dim D_2) such that $d(D_1,D_2) \leq \lambda$; the Banach -Mazur distance $d(D_1,D_2)$ is defined as $\inf\{\|T_i\|\cdot\|T_i^{-1}\|\}$, the lower bound is taken over all isomorphisms of D_1 onto D_2.)Then there exists an equivalent norm in B such that, denoting by B_o and D_o the spaces B and D equipped with the new norm, D_o admits a unique minimal projection from B_o.

Proof. The statement follows directly from Corollary 2.14 combined with the result of P. Enflö on the existence (in a superreflexive space) of an equivalent norm, which is simultanously uniformly convex and uniformly smooth (see e.g. [88],p.110).

Remark I.2.26. Let B be a finite-dimensional Banach space and D a subspace of B of codimension 1. Let $\rho(B,D) = p$. It can be shown that, for any $\varepsilon > 0$, the space B may be renormed so that the minimal projection onto the renormed subspace D_o be unique and its norm be strictly less than $p + \varepsilon$.

Problem I.2.27. Is the last assertion valid for a superreflexive infinite-dimensional space B and a subspace D of codimension 1?

———————— * ————————

We conclude this section with a result of V. M. Kadeč on a Lipschitzian projection with unit norm. Given a Banach space B and a subspace D, an idempotent, continuous operator $Tl: B \to D$ is called a Lipschitz projection if there is a number $\lambda > 0$ such that $\|Tl(x) - Tl(y)\| \leq \lambda\cdot\|x - y\|$ for all $x,y \in B$; we define

$$\|Tl\| = \inf\{\lambda > 0: \|Tl(x) - Tl(y)\| \leq \lambda\cdot\|x - y\| \text{ for all } x,y \in B\}.$$

For other related results see e.g. [138],[160].

Theorem I.2.28. **(Kadeč W.M; first published here, under the author's permission)** Let D be a subspace of a Banach space B and let $\Pi: B \to D$ be a Lipschtz projection with unit norm. Suppose the set

$$F_D = \{g(y,\cdot): y \in S_D \cap \text{sm } S_B\}$$

is total on D (see Intr., section 2 for the definition of $g(\cdot,\cdot)$). Then Π is a minimal projection and unique one.

Proof. Define $D_1 = \{x \in B: g(z,x) = 0 \text{ for all } z \in S_D \cap \text{sm } S_B\}$. By hypothesis, $D_1 \cap D = \{0\}$. We verify that Π is a linear operator anihilating D_1. To this end it suffices to show that

$$\Pi^{-1}(y) \subset y + D_1 \text{ for any } y \in D. \qquad (1.2.13)$$

Let $y \in D$, $z \in S_D \cap \text{sm } S_B$. Pick an arbitrary element $x \in \Pi^{-1}(y)$. To prove (1.2.13), it is enough to check that $g(z,x-y) = 0$. Take $t > 0$. Then $\|z\|/t = \|y - (y+z/t)\| = \|\Pi x - \Pi(y+z/t)\| \le \|x - y - z/t\|$,

$\|z\|/t = \|y - (y-z/t)\| = \|\Pi x - \Pi(y-z/t)\| \le \|x - y + z/t\|$.

Hence $(\|z-t\cdot(x-y)\| - \|z\|)/t \ge 0$, $(\|z-t\cdot(y-x\| - \|z\|)/t \ge 0$.

Since t was chosen arbitrarily, we get $g(z,x-y) \ge 0$ and $g(z,y-x) \ge 0$. But this means that $g(z,x-y) = 0$.***

Remark I.2.29. a) It is seen from the proof of the last theorem that if $B \ne D + D_1$ then there is no projection from B onto D with unit norm.

b) If F_D is a norming set on D (i.e. $\|x\| = \sup\{f(x): f \in F_D\}$ for every $x \in D$) then $D + D_1$ is a closed subset of B (a subspace, in fact) and a projection from $D + D_1$ onto D paralell to D_1 has unit norm. These two statements follow from the observation that

$$\|y\| = \sup\{f(y): f \in F_D\} = \sup\{f(y+y_1): f \in F_D\} \le \|y + y_1\|$$

for $y \in D$, $y_1 \in D_1$ (i.e. $D_1 \perp D$). (See also below, the statement a°) preciding Proposition 3.2 and the proof of Lemma 4.12.

§ 3. On the uniqueness or nonuniqueness of minimal projections in the case dim(B/D) ≥ 2

In the consideration of this section , devoted to the problem of uniqueness of minimal projections from a Banach space B onto a subspace D of codimension ≥ 2, the main role is played by the hypothesis on the existence of a 1-regular subspace $K \supset D$, i.e. a subspace admitting a projection of norm 1. In general, it can happen that there are no 1-regular subspaces of codimension ≥ 2 in B, as is shown by V.I. Gurarii's example

(see [175] for details). However, this does not occur in "good" situations.

And, thus we are going to inspect the diagram

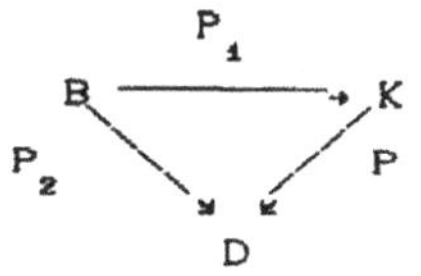

(1.3.1)

where $D \subset K \subset B$ and P, P_1, P_2 are projections, $\|P_1\| = 1$.

The following statements are easily verified:

Proposition I.3.1. Let D and K be subspaces of a Banach space B, $D \subset K \subset B$, and suppose that the subspace K is 1-regular. Then

1° if D is p-regular $(p \geq 1)$ in B, then D is p-regular in K; moreover, if $P \in \Delta^p(B,D)$, then $P|_K \in \Delta^p(K,D)$;

2° if D is p-regular $(p \geq 1)$ in K, then D is p-regular in B as well; moreover, if $P \in \Delta^p(B,D)$, $P_1 \in \Delta^1(B,K)$, then $P_2 = P \circ P_1 \in \Delta^p(B,D)$;

3° if card $\Delta(B,D) = 1$ then card $\Delta(K,D) = 1$;

4° if card $\Delta(B,D) = 1$, card $\Delta(B,K) > 1$ and there exist at least two distinct projections $P_{1(1)}$, $P_{1(2)} \in \Delta^1(B,K)$ such that $P^{-1}_{1(2)}(0) \subset (P^{-1}_{1(1)}(0) \oplus P^{-1}(0))$, where $P \in \Delta(B,D)$, then card $\Delta(B,D) > 1$.

We will need in the sequel three further propositions (3.2,3.3,3.4) on the uniqueness of norm one projection.

The proof of the first of them is based on Corollary 2.20 and the following facts, easy to verify:

a°) if $D \oplus K = B$, then the orthogonality relation $D \perp K$ is equivalent to the existence of a projection onto D parallel to K with unit norm;

b°) if $B = [D \oplus K]_p$, $1<p\leq\infty$, then $\mathcal{J}_x \supset K$ for any $x \in K \setminus \{0\}$; and moreover, $D \perp K$ and $K \perp D$.

The second proposition is checked in a straightforward way and the third is an easy consequence of the second. (All these three propositions are proved in the paper [140], Example 1.1, Theorem 3.4, Corollary 3.1).

Proposition I.3.2. Let K and D be a subspaces of a Banach space B, $D\neq\emptyset$, $K \neq \emptyset$, and suppose that $B = [K \oplus D]_p$, $1<p\leq\infty$. Then

$$\text{card } \Delta(B,K) = \text{card } \Delta(B,D) = 1.$$

Proof. Set $\hbar^D = D^*$, $\hbar^K = K^*$. Statements a°) and b°) (above) show that the pairs (B,K) and (B,D) satisfy all conditions of Corollary 2.20. (See also [113], Corollary 2.1).***

Proposition I.3.3. Let B be a Banach space and suppose that $B=[D \oplus [y]]_1$,

$y \notin D \neq \{0\}$. For any $x_o \in W_D = \{x \in D: 0 \le \|x\| \le 1\}$ define an operator T_{x_o} by

$$T_{x_o}(x + c\cdot y) = x + c\cdot\|y\|\cdot x_o \quad (x \in D,\ c \in \mathbb{R}). \tag{1.3.2}$$

Then T_{x_o} is a projection onto D with norm 1.

Proposition I.3.4. Let $B = [D \oplus K]_1$, $D \neq \emptyset$, $\dim K \ge 1$. For every $y \in K$ (also for $y = 0$)the subspace $[D \oplus [y]]_1$ is 1-regular in B, but card $\Delta(B,D) > 1$.

Corollary I.3.5. Let $B = [K \oplus X]_1$. Let D be a p-regular subspace of K, $p \ge 1$. Let $\dim X \neq 0$ and card $\Delta(K,D) = 1$. Then D is p-regular also in B, but card $\Delta(B,D) > 1$.

Proof. Choose an arbitrary element $y \in X \setminus \{0\}$. According to Proposition 3.4, subspaces K and $K_1 = [K \oplus [y]]_1$ are 1-regular in B.
Let $P \in \Delta^p(K,D)$. Choose $x_o \in D \subset K$, $x_o \neq 0$, and write $P_2 = T_{x_o}$, $P_3 = T_o$, where T_{x_o} and T_o are projections from K_1 onto K defined by formula (1.3.2). Clearly, $P_2 \neq P_3$. Since $P_2(-\|y\|\cdot x_o+y) = 0$ and $(P \circ P_3)(-\|y\|\cdot x_o+y) = -\|y\|\cdot x_o$, we see that the the point $-\|y\|\cdot x_o+y$ belongs to $P_2^{-1}(0)$ but not to $(P^{-1}(0)) \oplus P_3^{-1}(0))$, i.e. $P_2^{-1}(0)$ is not contained in $(P^{-1}(0) \oplus (P_3^{-1}(0))$.
Hence, by Proposition 3.1, statements 4° and 2°, card $\Delta(K_1,D) > 1$ and consequently card $\Delta(B,D) > 1$.***

*

Remark I.3.6. Diagram (1.3.1) can be employed not only to an examination of the uniqueness problem for a pair (B,D), but also in proving the lack of 1-regularity of subspaces containing the given subspace D.

Consider the following

Example I.3.7. The space $C_o(2\pi)$ does not contain a reflexive 1-regular subspace containing π_n, for any $n \ge 1$.

Proof. Assume it does; i.e. suppose there exists a 1-regular reflexive subspace $K \supset \pi_n$.

Let $P \in \Delta^1(C_o(2\pi),K)$. The Fourier projection F_n is minimal, and thus by Proposition 3.1 the restriction $F_n|_K$ has the same norm as F_n. Now, K being reflexive, projection $F_n|_K$ attains its norm on S_K (see Lemma 2.4). Thus also F_n attains its norm on $S_K \subset S_B$, in contradiction to Lemma 2.7. ***

Theorem I.3.8. Let K be a 1-regular subspace of B, let D be a p-regular ($p \geq 1$) subspace of K and suppose that card $\Delta(K,D) = 1$. Let $P_o \in \Delta(K,D)$.

A sufficient condition for the equality card $\Delta(B,D) = 1$ is the existence of a set $\hbar^D \subset S_{D^*}$ with the properties:

(i°) $\hbar^D$ is total on D;

(ii°) for any $f \in \hbar^D$ the functional $f \circ P_o$ has property (U) in B;

(ii°) $\hbar^D \subseteq E^K_P|_D$.

Proof. Assume that $P_1 \in \Delta^1(B,K)$, $P_3 \in \Delta^P(B,D)$. Write $P_2 = P_o \circ P_1$ and $\hat{P} = P_3|_K$. By Proposition 3.1, $P_2 \in \Delta^P(B,D)$, $\hat{P} \in \Delta^P(K,D)$. Since card $\Delta(K,D) = 1$, $\hat{P} = P_o$. Choose a functional $f \in \hbar^D$. By assumption (iii°), $f \in E^K_P|_D$ and so the functional $f \circ P_o$ has two norm-preserving extentions, namely, $f \circ P_o \circ P_1$ and $f \circ P_3$. They must be equal, according to assumption (ii°). The assertion of the theorem now follows from Theorem 2.19. ***

Corollary I.3.9. Let D be a finite-dimensional subspace of B and let $K \supset D$ be a 1-regular subspace of B. Suppose that card $\Delta(K,D) = 1$ and let $P \in \Delta(K,D)$. Further, suppose that one of the following conditions is satisfied:

a) K is reflexive and $K + \mathcal{J}_z = B$ holds for any $z \in$ crit P;

b) every functional of type $f \circ P$, where $f \in \bigcup_{y \in M^{-D}_P} A^K(y)$,

has property (U) in B.

Then card $\Delta(B,D) = 1$.

Proof. On account of Lemmas 2.1 ,2.4 and Proposition 2.16, the set $E^K_P|_D$ is total on D. Condition a) ensures that a functional of type $f \circ P$ has property (U) on B (uniqueness of norm preserving extention to all of B; see [164]). Hence, conditions of Theorem 3.8 are satisfied.***

Remark I.3.10. Property (U)-uniqueness of norm-preserving extention - turns out to be crucial in Theorem 3.8 as well as in Corollary 3.9. Let us mention in this connection P.K. Belobrow's paper [14]. There is introduced an operator π which associates with any functional $f \in S_{D^*}$ (where D is a subspace of B) the set of all functionals in S_{B^*} extending f, with norm preserved. The uniqueness property (U) corresponds to the univalance of the operator π. Various necessary and sufficient conditions for such univalence have been given in [14], and sooner yet in [162].

Corollary I.3.11. (Theorem 3.1 of [142]). Let D and K be a subspaces of a space B, p-regular and 1-regular respectively, and $D \subset K$. Let card $\Delta(K,D) = 1$ and $P \in \Delta(K,D)$. In order that card $\Delta(B,D) = 1$ it is sufficient that there exist sets $\hbar_P \subseteq$ crit P ($\hbar_P \neq \emptyset$) and $\hbar^D \subseteq \bigcup_{y \in P(\hbar_P)} (A^K(y)|_D)$ such that

1^{oo}) $\hbar^D$ is total on D;

2^{oo}) for any $f \in \hbar^D$ the functional $f \circ P$ has property (U) in B.

Proof. In view of Lemma 2.1, $\hbar^D \subset E^K_P|_D$, and hence the condition of Theorem 3.8 is fulfilled. (Let us again remark that condition 2^{oo} is certainly satisfied if the equality $K + \mathcal{G}_z = B$ holds for each $z \in \hbar_p$.***

Corollary I.3.12. Let D be a uniformly convex subspace of a smooth space K, dim K/D = 1. Let $P \in \Delta(K,D)$ and suppose that $E^K_P|_D$ is total on D (this is in the case, for instance, when dim d < ∞). Let X be a Banach space and consider $B = [K \oplus X]_p$ $(1 < p \leq \infty)$. Then card $\Delta(B,D) = 1$.

Proof. Note that K is a 1-regular subspace of $B = [K \oplus X]_p$, in virtue of Proposition 3.2. Moreover, for any $z \in D$ we have $X \subset \mathcal{G}_z$, and hence $\mathcal{G}_z = X \oplus K = B$.
Now, if $\|P\| = p = 1$, then the assertion follows from Corollary 3.11; if $\|P\| > 1$, the assertion follows from Theorems 2.13 and 3.8.***

Remark I.3.13. If $P \in \Delta(K,D)$, K is a Banach space, and if P does not attain its norm (which , of course, is possible only if $\|P\| > 1$), the verification of conditions (ii^{oo}) and (iii^{oo}) of Theorem 3.8 becomes cumbersome. Therefore we now give other sufficient conditions.

Proposition I.3.14. Let K and D be subspaces of a Banach space B, let K be 1-regular and D be p-regular (p > 1) in B. Let $K \supset D$, card $\Delta(K,D) = 1$, $P_o \in \Delta(K,D)$, $P_1 \in \Delta^1(B,K)$.

A sufficient condition for the equality card $\Delta(B,D) = 1$ is the existence of a set $\hbar^D \subseteq S_{D^*}$ with the properties

(i^{oo}) $\hbar^D$ is total on D;

(ii^{oo}) for any $f \in \hbar^D$ and any subspace $K_1 \subseteq B$, $K_1 \supset K$, dim $K_1/K = 1$, there exists a sequence $(z_m)_1^\infty$, $z_m \in S_{K_1}$ (m=1,2,...)such that

$$\inf (\|z_m - P_1 z_m\| : m \in \mathbb{N}) > 0; \tag{1.3.3}$$

$$\sup \{\|z_m - 2\cdot P_1 z_m\| : m \in \mathbb{N}\} \le 1; \qquad (1.3.4)$$

$$\lim_{n\to\infty} f((P_o \circ P_1)(z_m)) = p. \qquad (1.3.5)$$

Proof. Assume that there is an operator $P \in \Delta(B,D)$ other than $P_o \circ P_1$ (obviously, $P_o \circ P_1$ is in $\Delta(B,D)$). Since card $\Delta(K,B) = 1$, we have $P|_K = P_o$ and thus there exists $z \in S_B \cap P^{-1}(0)$ with $Pz \neq \emptyset$. Write $K_1 = K \oplus [z]$.

Let $f \in \hbar^D$. Evidently, $(1/p)\cdot f\circ P$ and $(1/p) f\circ P_o \circ P_1$ are norm-preserving extensions of the functional $(1/p)\cdot f\circ P_o$ from K to B. According to assumption (ii^{∞}), there exists a sequence $\langle z_m\rangle \subset S_{K_1}$ satisfying (1.3.3), (1.3.4) and (1.3.5). For each $m \in \mathbb{N}$ define $x_m = P_1 z_m$ and $\alpha_m = z_m - x_m$ ($\alpha \in \mathbb{R}$). Clearly the sequence $\langle \alpha_m \rangle_1^{\infty}$ is bounded and hence contains a convergent subsequence $\langle \alpha_{m_i} \rangle$ (for the sake of simplicity we do not change the indexes).

Let $\alpha_o = \lim_{m\to\infty} \alpha_m$. By (1.3.3) $|\alpha_o| > 0$. Assume that $(f\circ P)(z) \neq 0$. We distinguish two cases.

a) $\alpha_o \cdot (f\circ P_o)(z) > 0$. Then by (1.3.5)

$$1 = \|(1/p)\cdot(f\circ P)\| \ge \lim_{m\to\infty} |(1/p)\cdot(f\circ Pz_m)| = \lim_{m\to\infty} |(1/p)\cdot(f\circ P)(x_m + \alpha_m z)|$$

$$= |1 + (1/p)\cdot\alpha_o\cdot(f\circ P)(z)| > 1, \text{ a contradiction.}$$

b) $\alpha_o \cdot (f\circ P)(z) < 0$. Then, in view of (1.3.4),

$$1 \ge \|z_m - 2\cdot P_i z_m\| = \|x_m - \alpha_m \cdot z\|,$$

whence by (1.3.5)

$$1 = \|(1/p)\cdot(f\circ P)\| \ge \lim_{m\to\infty} |(1/p)\cdot(f\circ P)(x_m - \alpha_m\cdot z)| =$$

$$= |1 - (1/p)\cdot\alpha_o\cdot(f\circ P)(z)| > 1;$$

we have again arrived at a contradiction.

Therefore, $(f\circ P)(z) = 0$. The set $\hbar^D$ being total, we get $Pz = 0$, contrary to the choice of z.***

Remark I.3.15. To have condition (1.3.4) fulfilled, it suffices to require, for instance, that

$$\|x + y\| = \|x - y\| \qquad (1.3.6)$$

for every $x \in K$ and $y \in P^{-1}(0)$.

Corollary I.3.16. Let D be a p-regular ($p > 1$) subspace of K with card $\Delta(K,D) = 1$. Let X be a Banach space and consider $B = [K \oplus X]_{\infty}$ ($\dim X \ge 1$). Let $P \in \Delta(K,D)$. If the set $E_K^P|_D$ is contains a subset $\hbar^D$ total on D, then card $\Delta(B,D) = 1$.

Proof. First of all, observe that K is 1-regular in B and card $\Delta(B,K)=1$, according to Proposition 3.2. Let K_1 be a subspace of B with $K_1 \supset K$, $\dim K_1/K = 1$. Then clearly $K_1 = [K \oplus [x_o]]_{\infty}$ for a certain $x_o \in S_{K_1} \cap X$.

Let $f \in \hbar^D$. Then $f \in E^P_K|_D$ and thus there exists a sequence $(y_m) \subset S_K$ such that $\|(f \circ P)(y_m)\|$ tends to $\|P\| = p$, as $m \to \infty$. (Hence $f \in S_D*$). For each $m \in \mathbb{N}$ write $z_m = y_m + x_o$. It is not hard to check that the sequence $(z_m)_1^\infty$ satisfies conditions (1.3.3),(1.3.4) and (1.3.5). Thus, by Proposition 3.14, card $\Delta(B,D) = 1$.***

Corollary I.3.17. Let D be a p-regular ($p > 1$) subspace of K with $\dim D < \infty$ and card $\Delta(K,D) = 1$. Let X and B be as in Corollary 3.16. Then card $\Delta(B,D) = 1$.

Proof. Follows immediately from Corollary 3.16, Theorem 2.17 and Lemma 2.1.***

Example I.3.18. Let $K = L_1([-1;1])$, $D = \pi_1$, the subspace of L_1 consisting of all polynomials of degree ≤ 1. Let $B = [K \oplus X]_\infty$, where $\dim X \ge 1$. Then card $\Delta(B,D) = 1$.

To see this, first note that card $\Delta(K,D) = 1$, $\rho(K,D) = p > 1$.(See [44], Th.4). By virtue of Lemma 2.1 and Theorem 2.17, the set $E^K_P|_D$ (where $p \in \Delta(K,D)$) is total on D. Thus our assertion follows from Corollary 3.17.***

Example I.3.19. Let $B = [K \oplus \mathbb{R}]_\infty$, $K = C_o(2\pi)$, $D = \pi_n$, $n \ge 1$. Then K is 1-regular in B, D is p-regular in B ($p > 1$) and card $\Delta(B,K) =$ $=$ card $\Delta(B,D) = 1$.

It sufficies to remark that card $\Delta(C_o(2\pi),\pi_n) = 1$ (see formula (0.2.3) with the accompanying text) and the set $E^K_{P_3}|_D$ (where $P_3 \in \Delta(C_o,\pi_n)$) is total on D, on account of Lemma 2.1 and Theorem 2.17. The claim now follows from Proposition 3.2 and Corollary 3.17.

It is worth while noticing that the projection P in $\Delta(B,D)$ does not attain its norm. Indeed; let $P_1 \in \Delta(B,K)$. Taking into account that P_3 does not attain its norm (see Lemma 2.7) and that $P = P_3 \circ P_1$ (since card $\Delta(B,D) = 1$), we see that P cannot attain its norm, either.***

Remark I.3.20. It should be emphasized, in connection with the last example, that the condition card $\Delta(B,D) = 1$ not always provide the uniqueness of a minimal projection (given that it exists) onto a subspace containing D. Consider, for example, the diagram

$$\begin{array}{ccc} B = [C_o(2\pi) \oplus \mathbb{R}]_\infty & \xrightarrow{\;P_2\;} & D_1 = [\pi_n \oplus \mathbb{R}]_\infty \\ {\scriptstyle P_3}\big\downarrow & \swarrow{\scriptstyle P_4} & \\ D = \pi_n & & \end{array} \qquad (1.3.7)$$

Let $P \in \Delta(B,D)$, $P_4 \in \Delta(D_1,D)$, P_2 $\Delta(B,D)$ ($\|P\| > 1$, $\|P_4\| = 1$). Observe that card $\Delta(B,D)$ = card $\Delta(D_1,D) = 1$. By Proposition 3.1 D_1 is a p-regular subspace of B; nevertheless card $\Delta(D_1,D) > 1$.***

Example I.3.21. Let D and K be finite-dimensional subspaces of l_p (spaces isometric to subspaces of l_p, to be precise) ($1< p< +\infty$) with $K \supset D$, dim $K/D = 1$, dim $D \geq 2$. Let X be a space and let $B = [K \oplus X]_t$ ($1<t <+\infty$). Then card $\Delta(B,D) = 1$.

It is enough to note that D is uniformly convex (= strictly normed, since dim $D < \infty$) and K is smooth, according to general properties of l_p spaces ($1 < p < \infty$). The statement results by Corollary 3.12.***

Example I.3.22. Let D be a two dimensional subspace of a space K, dim K = 3, $P \in \Delta(K,D)$, $\|P\| > 1$. Let X be a Banach space and let B = $= [K \oplus X]_p$ ($1 < p \leq \infty$). Then card $\Delta(B,D) = 1$.

Just notice that every functional in K^* extends to B with norm unchanged, because $\mathcal{J}_z \supset X$ for any $z \in S_K$ (see statement b°) preceding Proposition 3.2) it remains to appeal to Theorem 1.3 and 3.8.***

§ 4. Strong uniqueness of minimal projections

Let D be a complemented subspace of a Banach space B, dim $B/D \geq 2$. In this section we examine the problem of existence and uniqueness of a minimal projection from B onto D under the condition that card $\Delta(K,D) =1$ for any $K \supset D$ with dim $K/D = 1$. The (unique) projection in $\Delta(K,D)$ will be denoted by P_K. If card $\Delta(B,D) = 1$ and card $\Delta(K,D) = 1$ for every subspace $K \subset B$ with dim $K/D = 1$, the unique projection in $\Delta(B,D)$ will be called **strongly unique.**

We also give in this section a necessary and sufficient conditions for the existence (and uniqueness) of a minimal projection onto a codimension one subspace, condition originating in the investigations of papers [23],[24],[73],[74],[75],[122].

Definition I.4.1. Let D be a subspace of B and let $x \in D$. The set

$$\bigcap_{y \in D} W_D(y;\rho(B,D)\cdot\|x-y\|) \tag{1.4.1}$$

will be called the set of minimal points fo x and D and will be denoted by $B_D(x)$.

If $x \in D$ then clearly $B_D(x) = \{x\}$.

Lemma I.4.2. Let D be a subspace of a Banach space B, dim B/D = 1, and let $x \in B \setminus D$. Suppose $B_D(x) \neq \emptyset$. Then, given any point $z \in B \setminus D$, $z = \alpha\cdot x + y_z$ ($\alpha \in \mathbb{R}, y_z \in D$), we have the set equality

$$B_D(z) = \alpha\cdot B_D(x) + y_z.$$

Proof. Let $y_o \in B_D(x)$. We show that $\alpha\cdot y_o + y_z \in B_d(z)$. Note that $\alpha \neq 0$ (otherwise $z \in D$). Since $y_o \in B_D(x)$, we have $\|y_o - y\| \leq \rho(B,D)\cdot\|x-y\|$ for all $y \in D$ and thus

$$\|\alpha\cdot y_o + y_z - y\| = \|y_o + (1/\alpha)\cdot y_z - (1/\alpha)\cdot y\|\cdot|\alpha| \leq$$

$$|\alpha|\cdot\rho(B,D)\cdot\|x + (1/\alpha)\cdot y_z - (1/\alpha)\cdot y\| = \rho(B,D)\cdot\|\alpha\cdot x + y_z - y\| =$$

$$\rho(B,D)\cdot\|z - y\|,$$

i.e. $\alpha\cdot y_o + y_z \in B_D(x)$.

Now let $y_o \in B_D(z)$. An analogous argument shows that $(1/\alpha)\cdot(y_o - y_z) \in B_D(x)$.***

Corollary I.4.3. Let D be a subspace of of B, dim B\D = 1 and let $x \in B \setminus D$. If card $B_D(x) = 1$, then card $B_D(z) = 1$ for every $z \in B \setminus D$.

Proposition I.4.4. Let D be a subspace of a Banach space B, dim B\D = 1. A minimal projection onto D exists if and only if there is an element $x \in B \setminus D$, for wchich the set $B_D(x)$ is not empty.

Proof. Let P be a minimal projection onto D and let $\|P\| = \rho(B,D) \geq 1$. (For $\rho(B,D) = 1$ the proof has been in fact given by G. Godini ([73],Th.2). (See also [122])) Then, choosing any $x \in B \setminus D$, $y \in D$, we have

$$\|Px - y\| = \|P(x - y)\| \leq \|P\|\cdot\|x - y\| = \rho(B,D)\cdot\|x - y\|$$

and hence $Px \in B_D(x)$, so that $B_D(x) \neq \emptyset$.

Now suppose that $B_D(x) \neq \emptyset$ for some $x \in B \setminus D$. Let $y_1 \in B_D(x)$. Every element $z \in B$ can be uniquely written as a sum $z = \alpha\cdot x + y_z$ with $\alpha \in \mathbb{R}$, $y_z \in D$. Define $Pz = \alpha\cdot y_1 + y_z$ for $z = \alpha\cdot x + y_z$.

That P is a projection, it is evident. On account of Lemma 4.2 we have

$$\|P\| = \sup(\|Pz\|: \|z\| = 1) \leq \sup(\|\alpha\cdot y_1 + y_z\|: \|\alpha\cdot x + y_z\| = 1) \leq$$

$$\leq \rho(B,D)\cdot\sup(\|\alpha\cdot x + y_z\|: \|\alpha\cdot x + y_z\| = 1) = \rho(B,D).$$

Since P projects onto D, $\|P\| \geq \rho(B,D)$; thus $\|P\| = \rho(B,D)$.***

Remark I.4.5. In the formulation of Proposition 4.4 we have demanded the existence of at least one element $x \in B \setminus D$ for which $B_D(x) \neq \emptyset$. However, it follows from Lemma 4.2 (and from the proof of Proposition 4.4, as well) that if $B_D(x) \neq \emptyset$ for some for some $x \in B \setminus D$ then actually

$B_D(z) \neq \emptyset$ for all $z \in B$.

Proposition 4.4 and Corollary 4.3 result in

Corollary I.4.6. Let D be a subspace of B, dim B/D = 1. A necessary and sufficient condition for the existence and uniqueness of minimal projection onto D is given by:

$\operatorname{card} B_D(x) = 1$ for every $x \in B \setminus D$.

Theorem I.4.7. Let $D \subset B$ and assume that, given any subspace $K \subseteq B$ with $K \supset D$, $\dim K/D = 1$, we have $\operatorname{card} \Delta(K,D) = 1$ and $P_K \in \Delta(K,D)$. Define

$$Px = \begin{cases} P_K x \text{ if } x \in B \setminus D\ ,\ K = D \oplus [x]; \\ x \quad \text{if } x \in D. \end{cases} \tag{1.4.4}$$

The additivity of operator P is a sufficient condition for the existence of of a minimal projection from B onto D; this is also a necessary condition in the case when $\|P_K\| = \rho(B,D)$ for every $K \supset D$ with $\dim K/D = 1$.

Moreover, under this condition, the minimal projection onto D is strongly unique (and coincides with P).

Proof 1. Suppose that the operator P given by (1.4.4) is additive. It is also homogenous, because so is each P_K. The idempotency of P is evident. Hence, P is a projection onto D.

Assume that the projection P is not minimal. Thus there exists a projection $P_1 \in \Delta(B,D)$, $P_1 \neq P$, $\|P_1\| < \|P\|$. Consequently there is an element $x \in S_B$ such that $\|Px\| > \|P_1\|$. Clearly, $x \notin D$ (else $\|Px\|=1\le\|P_1\|$). Write $K = D \oplus [x]$. Since $P|_K = P_K$, we have

$$\|P|_K\| = \|P_K\| \ge \|P_K x\| > \|P_1\| \ge \|P_1|_K\|,$$

showing that P_K is not a minimal projection from K onto D, contrary to the condition of the theorem. This contradiction means that P is a minimal projection - a unique one, by definition.

2. Now suppose P_1 is a minimal projection of B onto D. Then $\|P_1\|=\rho(B,D)$. Assume that $\|P_K\| = \rho(B,D)$ for every $K \supset D$, $\dim K/D = 1$. Since

$$\rho(B,D) = \|P_1\| = \|P_1|_K\| \ge \|P_K\| = \rho(K,D) = \rho(B,D),$$

$P_1|_K$ is a minimal projection from K onto D, and hence $P_1|_K = P_K$, i.e. $P_1 = P$. Then, of course, P is additive.***

Proposition I.4.8. Let D be a two-dimensional subspace in B, dim B ≥ 4, admitting a unique minimal projection from B. Suppose that every three-dimensional subspace which contains B is smooth. Then the minimal projection from B onto D is strongly unique.

Proof. It suffices to observe that, given any subspace $K \supset D$, $\dim K/D = 1$, the minimal projection is strongly unique. Now, if $\|P_K\| > 1$, then $\operatorname{card} \Delta(K,D) = 1$ by Theorem 1.7; if $\|P_K\| = 1$, then $\operatorname{card} \Delta(K,D) = 1$ by the smoothness of K (see Corollary 2.20 or Corollary 1 of [116]).***

Example I.4.9. Consider the space $B = [K \oplus X]_{p_2}$ and its subspace X, where $K = l^3_{p_1}$, $p_1 > 2$, $1 < p_2 < \infty$. Let D be a p-regular subspace of K, $p > 1$. Such a subspace exists, since $p_1 > 2$ (see below, Proposition III.3.6). The minimal projection from B to D has norm p and it is strongly unique, because each subspace of dimension 3 containing D is smooth (see statement b°) foregoing Proposition 3.2) and Proposition 4.8 applies.***

---------------- * ----------------

Let D be a subspace in B and suppose that for any $x \in B \setminus D$ there exists a unique minimal projection $P_x : D \oplus [x] \to D$. Write

$$Ort_D(x) = \{z \in S_D : z \perp P_x^{-1}(0)\}.$$

We will denote by χ^D a selector defined on S_D with values in $\bigcup_{z \in D} A^B(z)$, so that $\chi^D(z) \in A^B(z)$.

Definition I.4.10. A subspace D is said to have the intersection property in B if, given any three elements x_1, x_2, $x_3 \in B \setminus D$, the set

$$\mathcal{F}(x_1, x_2, x_3) = \bigcap_{i=1}^{3} Ort_D(x_i)$$

is nonempty.

D is said to have the strong intersection property in B if, given any three elements x_1, x_2, $x_3 \in B \setminus D$ and an arbitrary selektor χ^D,

(i) $\mathcal{F}(x_1, x_2, x_3) \neq \emptyset$;

(ii) $\bigcup_{x \in \mathcal{F}(x_1, x_2, x_3)} (\chi^D(x)|_D)$ is total on D.

Theorem I.4.11. Let $D \subset B$ be a subspace having the strong intersection property in B. Suppose that card $\Delta(K, D) = 1$ for every subspace $K \supset B$, $\dim K/D = 1$. Let P be the operator defined by formula (1.4.4). Further, suppose that, given any $x, y \in B \setminus D$ such that $x + Y \in B \setminus D$, at least two of the three points: $x - Px$, $y - Py$, $(x + y) - P(x + y)$ belongs to $\mathcal{J}_z$, for each $z \in \mathcal{F}(x, y, x+y)$.

Then the minimal projection from B onto D exists and it is strongly unique.

Proof. If $x, y \in B$ are such that, at last one of the three elements: x, y, x+y belongs to D, then using the uniqueness of minimal projection onto D from each $K \supset D$ with $\dim K/D = 1$ it is easy to see that $P(x+y) = Px + Py$.

The additivity of operator P in case where $x, y, x+y \in B \setminus D$ (and hence, by Theorem 4.7, the strong uniqueness of P), is a consequence of the strong intersection property and the following lemma:

Lemma I.4.12. Let D be a subspace of B and let $x, y \in B$. Let $\mathcal{P}$ be an idempotent operator from B onto D. Let $\mathcal{E} \subseteq S_D$ be such that for any $z \in \mathcal{E}$,

1° $z \perp (x - \mathcal{P}(x))$, $z \perp (y - \mathcal{P}(y))$, $z \perp ((x+y) - \mathcal{P}(x+y))$;

2° of the three elements: $x - \mathcal{P}(x)$, $y - \mathcal{P}(y)$, $(x+y) - \mathcal{P}(x+y)$ at least two belongs to $\mathcal{G}_z$.

Then

$$z \perp (\mathcal{P}(x+y) - \mathcal{P}(x) - \mathcal{P}(y)); \tag{1.4.5}$$

$$(x+y) = \mathcal{P}(x) + \mathcal{P}(y), \tag{1.4.6}$$

provided that either, for each selector χ^D, the set $\bigcup_{z\in\mathcal{E}} (\chi^D(z))|_D$ is total on D, or $x, y \in D$.

Proof. First of all observe that, whenever w_1, w_2 are elements satisfying $g(z,w_1) = g(z,w_2) = 0$, then also $g(z,w_1+w_2) = 0$, by the linearity of the functional $g(z,\cdot)$ on $\mathcal{G}_z$ (see [164]). Further, according to the well known result of James [84], $g(z,w) = 0$ iff $z \perp w$ and $w \in \mathcal{G}_z$. Finally, we show that $g(z,y) = 0$ and $z \perp w$ yield $z \perp (y+w)$. To see this, consider any supporting hyperplane $K_v = z + M^z_v$ to the ball $W_B(0;\|z\|)$ such that $M^z_v \supset [w]$. (Recall that the supporting subspace to the ball $W_B(0,\|z\|)$ at a point $z \in B$ is defined as a subspace $M^z \subseteq \{y \in B: z \perp y\}$ such that $\dim B/M_z = 1$. M^z can be regarded (see [84]) to be "generated" by a functional from $A^B(z)$; the set $z+M^z$ is called a supporting hyperplane to $W_B(0;z)$; tangent hyperplane when $M^z = \{y \in B: z \perp y\}$.) Let $y \neq 0$, $w \neq 0$. Looking at a two-dimensional space $[z,y]$, we have $M^z_v \cap [z,y] = [y]$, since z is a point of smoothness in $[z,y]$, and therefore the supporting hyperplane to the ball $W_{[z,y]}(0;\|z\|)$ is in fact a tangent one. Since $z \perp M^z_v$, $(z+x) \perp y$.

In view of this remark, the ortogonality relation (1.4.5) follows immediately from the assumptions of the lemma.***

We pass to the proof of (1.4.6). If $x, y \in D$, (1.4.6) holds because $\mathcal{P}$ is idempotent. Assume that $[x,y] \cap D \neq [x,y]$. To any $z \in \mathcal{E}$ we can find, in view of (1.4.5), a subspace M^z supporting to W_B at z and containing $(\mathcal{P}(x+y) - \mathcal{P}(x) - \mathcal{P}(y))$. Let χ^D be such that $M^z = f^{-1}(0)$, where $f = \chi^D(z)$ for each $z \in \mathcal{E}$. Since $(\chi^D(z)\cdot(\mathcal{P}(x+y) - \mathcal{P}(x) - \mathcal{P}(y)) = 0$ and the set $\bigcup_{z\in\mathcal{E}} (\chi^D(z)|_D)$ is total on D, we have $\mathcal{P}(x+y) = \mathcal{P}(x) + \mathcal{P}(y)$.***

Remark I.4.13. To meet the requirements of Lemma 4.12 (conditions $1^\circ$and

2°) it suffices e.g. to demand that

$$g(z,x-\mathcal{K}(x)) = g(z,y - \mathcal{K}(y)) =g(z,(x+y) - \mathcal{K}(x+y)) = 0,$$

for then

$$g(z,\mathcal{K}(x+y) - \mathcal{K}(x) - \mathcal{K}(y)) = 0.$$

Instead of assuming that $\bigcup_{z\in\mathcal{E}} (\chi^{D}(z)|_{D})$ is total on D for arbitrarily chosen selector χ^{D}, we might have required that $\bigcup_{z\in\mathcal{I}?} (A^{B}(z)|_{D})$ be total because if $\chi^{D}(z) \in A^{B}(z)$ and $g(z,y)$ exists, then $(\chi^{D}(z))(y) = g(z,y)$. ***

Corollary I.4.14. Suppose D is a subspace of B, $\rho(B,D) = 1$. Then the following condition is sufficient for the existence and strong uniqueness of a norm 1 projection onto D:

For arbitrary $x_1, x_2 \in B \setminus D$,

$1^{\circ\circ}$ $B_D(x_i) \neq \emptyset$ for $i=1,2$;

$2^{\circ\circ}$ there exists a set $M_{x_1,x_2} \subseteq S_D \cap \mathrm{sm}\, S_{D\oplus[x_1,x_2]}$ such that

$$\bigcup_{z\in M_{x_1,x_2}} (A^{B}(z)|_{D}) \text{ is total on D.}$$

Proof. Let $x \in B \setminus D$. By Proposition 4.4, D is a 1-regular subspace of $K = D \oplus [x]$.

Condition $2^{\circ\circ}$ card $\Delta(K,D) = 1$ holds in view of Corollary 2.20. Subspace D has the strong intersection property in B. Hence, according to Remark 4.13, the minimal projection onto D is strongly unique. ***

Remark I.4.15. To fulfil condition $2^{\circ\circ}$ of Corollary 4.14, it is enough that the set $\bigcup_{x\in S_D\cap \mathrm{sm} S_B} g(x,\cdot)$ be total on D; and this is in the case when e.g. D has a Schauder basis, which is contained in $S_D \cap \mathrm{sm}\, S_B$.

If D is separable or reflexive then, in order to have condition $2^{\circ\circ}$ fulfilled, it suffices to require that the following equality (1.4.7) be satisfied in $M_{x_1,x_2} = D \oplus [x_1,x_2]$ for any $x_1,x_2 \in B \setminus D$:

$$D + \bigcap_{z\in S_D} \mathcal{I}_z = M_{x_1,x_2} \qquad (1.4.7)$$

(see e.g. Theorem 5.9.8 of Dunford, Schwartz' book [60] and Corollary 1 in [121]).

Instead of (1.4.7), the condition,

$$D + \bigcap_{z\in S_D} \mathcal{I}_z + B \qquad (1.4.8)$$

can be easier to verify.

In G.Godini's paper [73] (Th.2) condition $1^{\circ\circ}$ appears in the same form as in our Corollary 4.14, accompanied by

$$\mathrm{sm}\, S_D \cap \mathrm{sm}\, S_B = S_D \qquad (1.4.9)$$

instead of 2^{oo}. Obviously, (1.4.9) is a specific case of condition 2^{oo}. Consider e.g. the situation where D is the subspace in c_o spanned by the firs n elements ($n \geq 2$) of the natural basis in (c_o). (See below, Example 4.22). D is a 1-regular subspace of (c_o); the pair (c_o,D) satisfies condition (1.4.8), but not (1.4.9).***

Remark I.4.16. In [55] I.K.Daugavet has given necessary and sufficient conditions for the existence and uniqueness of norm 1 projection onto a finite-dimensional subspace of $C_{\mathbb{R}}(T)$, the space of all continuous real-valued functions on a compact metric space T, with the usual sup-norm. We will employ Corollary 4.14 to give a sufficient condition for the strong uniqueness of projection with unit norm onto a finite-dimensional subspace of $C_{\mathbb{R}}(T)$.

Thus let C_m be a subspace of $C_{\mathbb{R}}(T)$, $\dim C_m = m$, $m \geq 1$. Suppose C_m is spanned by functions $\phi_1,\dots,\phi_m$. Following [55] we write

$$\Phi(x) = (\phi_1(x),\dots,\phi_m(x)) \in \mathbb{R}^n,$$

$$\Phi = \{y \in \mathbb{R}^m : y = \Phi(x),\ x \in T\},$$

$$\Omega = \mathrm{conv}(\Phi \cup (-\Phi))$$

(for a set $E \subset B$, $\mathrm{conv}\, E = \{\sum_{i=1}^{n} a_i \cdot x_i : \sum_{i=1}^{n} a_i = 1,\ 0 \leq a_i \leq 1,\ x_i \in E\}$) and we call Ω an m-dimensional octahedron in $\mathbb{R}^m$ if Ω is a convex body in $\mathbb{R}^m$ with symmetry centre at the origin, having exactly m pairs of extremal points.

Proposition I.4.17. The following conditions are (jointly) sufficient for the existence and strong uniqueness of projection onto C_m with norm 1:

1^{ooo} Ω is an m-dimensional octahedron in $\mathbb{R}^m$;

2^{ooo} C_m contains m linearly independent peak functions.

Proof. We show that C_m satisfies condition 1^{oo} and 2^{oo} of Corollary 4.14. By virtue of the results of Daugavet [55] and Proposition 4.4, condition 1^{ooo} is equivalent to 1^{oo}. Further, every peak function in C_m is a smoothness point of $C_{\mathbb{R}}(T)$ (see [8], p.144, and also [143], Corollary 3.7). Consider linearly independent peak functions $x_1,\dots x_m \in C_m$; the set $\Omega = \bigcup_{i=1}^{m} g(x_i/\|x_i\|,\cdot)$ is total on C_m.***

Example I.4.18. Take a partition $0 = t_o < t_1 < \cdots < t_m = 1$ of the segment $[0,1]$ into m pieces. Let $\{x_i\}_{i=1}^{m}$ be a sequence of peak functions from $C_{\mathbb{R}}([0,1])$, each x_i attaining its norm at a point of the segment $[t_{i-1},t_i]$ and vanishing outside it. Then the space C_m spanned by $x_1,\ldots,x_m$ is 1-regular in $C_{\mathbb{R}}([0,1])$ and the minimal projection from $C_{\mathbb{R}}([0,1])$ onto C_m is strongly unique.***

*

In the remaining part of this section we construct an example of a pair (B,D) such that $\Delta(B,D) \neq \emptyset$, D has the strong intersection property in B, projection $P \in \Delta(B,D)$ is strongly unique and $\|P\| > 1$.

Lemma I.4.19. Let D be a two-dimensional subspace of a three-dimensional space B. Let $P \in \Delta(B,D)$. Then there are at least 6 elements $x_1, \ldots, x_6 \in S_D$ such that $x_i \perp P^{-1}(0)$ $(i=1,\ldots,6)$.

Proof. If $\|P\| = 1$, the claim is obvious, for then each $x \in S_D$ satisfies $x \perp P^{-1}(0)$. Thus let $\|P\| > 1$. Then the set $P(\mathrm{crit}\ P)$ has not less than 6 points (see Lemma 1.1) and we may assume, without loss of generality (see the proof of Theorem 1.3), that z_1^+, z_2^-, z_3^+, z_4^-, z_5^+, z_6^- are elements of $P(\mathrm{crit}\ P)$ labelled so that $z_i = -z_{i+3}$ $(i=1,2,3)$, their arrangement on $S_D(0;\|P\|)$ coinciding with the order they are listed above; those marked with a plus, resp. minus sign, being images of points from the one, resp. the other of the two halves into which B is partitioned by D.

Let $z_i \in S_B$, $Pz_i = z_i^+$ $(i=1,3,5)$. Let W_B^+ be the portion of W_B lying in the same half-space as z_1 and let W_B^- denote the other half. Write $y = -(Pz_1 - z_1)/\|(Pz_1 - z_1)\|$, $v_i^+ = z_i^+/\|P\|$, $v_j^- = z_j^-/\|P\|$ $(i=1,3,5; j=2,4,6)$. Then clearly $\|v_i^+\| = \|v_j^-\| = 1$, $\|y\| = 1$, $y \in P^{-1}(0)$. For small values of $t > 0$ we have $\|v_1^+ + t\cdot y\| < \|v_1^+\| = 1$, since $v_1^+ \in \mathrm{Int}\ W_B^+$ for $t > 0$ small enough.(See Fig. 9)

The arrangement of points z_1^+, z_2^-, z_3^+,... defines an orientation of the boundary $S_D(0;\|P\|)$. Let u_1 denotes the arc of S_D $(=S_D(0;1))$ with endpoints v_1^+ and v_2^- with the induced orientation. Let $\mathrm{Tr}:[0,1] \to S_D$ be a parametrization of u_1, i.e. a continuous map such that $u_1 = \mathrm{Tr}([0,1])$, $\mathrm{Tr}(0) = v_1^+$, $\mathrm{Tr}(1) = v_2^-$. Consider the set

$\omega_1 = \{x \in S_D : x + \lambda\cdot y \in \mathrm{Int}\ W_B^+,\ \lambda \in (0,1)\}$.

Clearly, $\omega_1 \neq \emptyset$, because $v_1^+ \in \omega_1$. If $x = \mathrm{Tr}(\nu_o) \in \omega_1$ for $\nu > \nu_o$, sufficiently closed to ν_o. Let

$\omega_1^+ = \{x \in \omega_1 \setminus \{v_1^+\} : x = \mathrm{Tr}(\nu), \mathrm{Tr}((0,\nu)) \subset \omega_1\}$.

Write $\alpha_o = \sup\{\nu : \mathrm{Tr}(\nu) \in \omega_1^+\}$. We have $\alpha_o < 1$, because $\mathrm{Tr}(1) = v_2^-$ and

$Tr(\nu) \notin \omega_1^+$ for ν close to 1. Let $x_1 = Tr(\alpha_0)$; this is one of the desired points. Indeed, $x_1 + t\cdot y \notin \operatorname{Int} \bar{W}_B$ (for any $t \in \mathbb{R}$), and so $x_1 \perp y$. Quite analogously, considering the remaining five arcs $u_2,\dots,u_6$ into which $S_D(0,1)$ is partitioned by points $z_i^{+,-}$, we find five other points on S_D, all of them orthogonal to y. ***

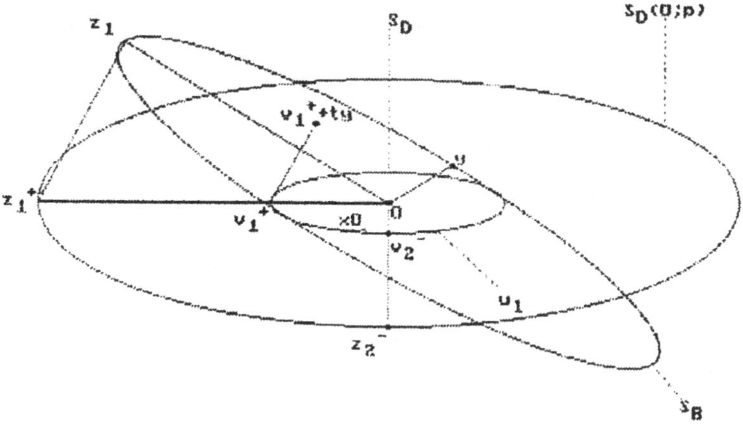

Figure 9

Corollary I.4.20. Let D be a two-dimensional subspace of a three-dimensional space B. Let $P \in \Delta(B,D)$, $y \in P^{-1}(0) \cap S_B$. Let $Ort_D(y) = \{x \in S_D : x \perp y\}$. Then

a^{oo}) the set $\bigcup_{z\in Ort_D(y)} (A^B(z)|_D)$ is total on D;

b^{oo}) if $[y] \subset \bigcap_{x\in S_D} \mathcal{J}_x$ then there exist at least 6 points $x_1,\dots,x_6 \in S_D$ with $g(x_i,y) = 0, i=1,\dots,6$.

Proof. Both assertions are obvious if $\|P\| = 1$. Thus let $\|P\| > 1$. By Lemma 4.19, there are at least 6 points on S_D such that $x_i \perp y$ $(i=1,\dots,6)$. Moreover, examining the proof of the lemma we see that at least two of these points do not lie on two parallel supporting hyperplanes of W_D. (the points z_1^+, z_2^-, z_3^+ in the proof of Lemma 4.19 do not lie on a common straight line; if they do, then necessarily $\|P\| = 1$). Hence, there exist two independent functionals corresponding to supporting hyperplanes at points of $Ort_D(y)$. Consequently, the set $\bigcup_{z\in Ort_D(y)} (A^B(z)|_D)$ is total on D.

Assertion b^{oo}) is a direct consequence of Lemma 4.19 and the fact that, whenever $g(x,y)$ exists, we have: $g(x,y) = 0$ iff $x \perp y$ (see [143]). ***

Proposition I.4.21. Let D be a two-dimensional subspace of an arbitrary Banach space B. Suppose that W_D has exactly 6 extremal points.(i.e. W_D is a convex hexagon). Further, suppose that card $\Delta(K,D) = 1$ for any $K \subset B$, $K \supset D$, $\dim K = 3$. Then D has the strong intersection property in B.

Proof. Let $x \in B \setminus D$. Write $K_x = D \oplus [x]$, choose $P_x \in \Delta(K_x,D)$ and let $x_o \in P_x^{-1}(0) \cap S_{K_x}$. We claim that $Ort(x_o) \supset ext\ W_D$. In the case of $\|P_x\|=1$ this is evident.

So let $\|P_x\| > 1$. Denote by $x_1,\dots,x_6$ the vertices of W_D numbered successively. According to Lemma 4.19 there exist 6 points $z_1,\dots,z_6 \in S_D$, numbered in agreement with the orientation of S_D induced by the x_i's, such that $z_i \perp P_x^{-1}(0)$ $(i=1,\dots,6)$ and, moreover the set $\langle z_i \rangle_{i=1}^{6}$ is not contained in two parallel lines supporting W_D (see the proof of Corollary 4.20). Therefore the points z_1, z_2, z_3 belong to the adjacent sides of S_D, at least.
Let z_1, $z_2 \in [x_1,x_2]$, with no less of generality. Then either $z_3 = x_3$ or $z_3 \in Int\ [x_2,x_3]$. (See Fig.10).

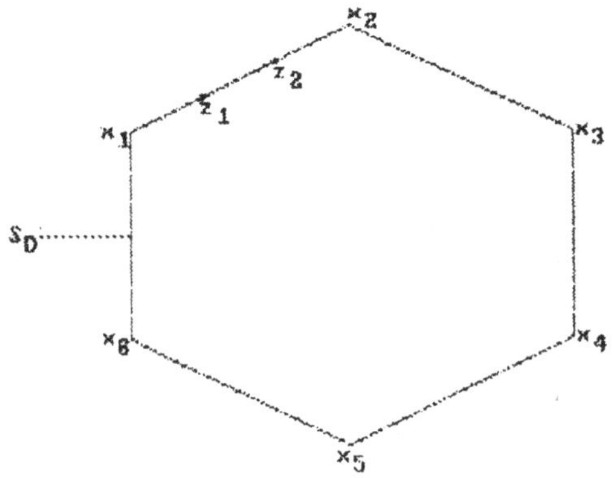

Figure 10

In each case it is easy to see that $Ort_D(x_o)$ contains three points x_1, x_2,x_3, and hence also the remaining vertices of S_D ; i.e. $x_i \perp P_x^{-1}(0)$ $(i=1,\dots,6)$, as claimed.

To conclude the proof , we observe that for any selector χ^D the set $\bigcup_{i=1}^{6}(\chi^D(x_i)|_D)$ is total on D. ***

Example I.4.22. Let B be either c_o or l^n_∞)(n≥3)and let $\langle e_i \rangle$ be the natural basis in B, $e_i = (0,\dots,0,\underset{i}{1},0,\dots)$. Consider the subspace of B spanned by h_1, h_2, where $h_1 = e_1 + e_2$, $h_3 = e_3 - e_1$. (See Fig. 11).

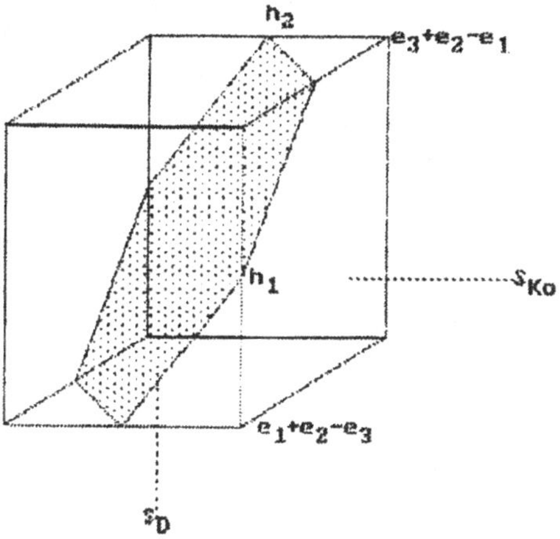

Figure 11

A minimal projection onto D exists and is strongly unique. Moreover, the space D has the strong intersection property in B.

First of all notice that, given any three-dimensional subspace K ⊆ B containing D, a minimal projection from K onto D is certainly unique; this follows from Theorem 1.3, in the case of norm > 1, and from Corollary 2.20, in case of norm = 1(because we can find a normalized basis in D whose elements are smoothness points of K; see also [144],Corollary 1). By Proposition 4.21, D has the strong intersection property in B.

Now, let $K_o = [e_1, e_2, e_3]$. Then card $\Delta(B, K_o)$ = card $\Delta(K_o, D) = 1$ and K_o is 1-regular in B. Let $\hbar = \langle e_3 + e_2 - e_1,\ e_1 + e_2 - e_3 \rangle \subset S_B$ and $\hbar^D = \langle e_1^*, e_2^* \rangle$, where $e_1^* = (-1,0,\dots)$ and $e_2^* = (0,1,0,\dots) \in S_{B^*}$. In view of Corollary 4.20, the triple (B, K_o, D) satisfies the conditions of Corollary 3.11. Thus the minimal projection from B onto D is unique, and hence also strongly unique. Its norm is equal to 4/3.(See also Section 3 of Chapter II).***

Remark I.4.23. Condition:
card $\Delta(K, D) = 1$ for any $K \subseteq B$ with $K \supseteq D$, dim $K/D = 1$ in Proposition 4.21 can be replaced by:

$$D + \bigcap_{y \in S_D} \mathcal{I}_y = B.$$

The assertion of Proposition 4.21 then remains valid, moreover, under this condition the minimal projection onto D certainly exists, its norm equals to 1 and the projection itself is strongly unique. This should be compared with the well known theorem of Lindenstrauss ([96], Th.2) on norm-preserving extensions of operators.

For the proof of our claim, observe that already in a three-dimensional space $K \supset D$ the conditions:

a) W_D has exactly 6 extremal points, and

b) $D + \bigcap_{y \in S_D} \mathcal{I}_y = K$

imply 1-regularity of D in K. The claim now results, in virtue of Corollary 4.14 and Remark 4.15. ***

Notes and remarks

1. Theorem I.1.3 was obtained by the first author in 1976 and published in 1978 in [145]. Proposition I.1.5 was a replay to the first author's problem, posed in [145]. K. Kürsten obtained this result in 1978, but did not publish it; he communicated it to the first author in January, 1980.

2. Theorems I.2.8 and I.2.13 were proved by the first author in 1974. (See [142]). Corollary I.2.20 has its origin in the works by A.E. Taylor [186] (1936), M.Z. Solomjak [182] (1956) and W. Wulbert [190] (1968). (See also [50] and [144]). Lemma I.2.1 first appeared in [156]. Lemma I.2.7 was mentioned in [146] and proved in [149] (1981). Theorem I.2.17 was first published in [156]; Theorem I.2.19 - in [149] Theorem I.2.28 was communicated to the first author by W.M. Kadeč in 1983. Some of the examples are taken from [151].(let us remark that the proof of the main result in [151] is not correct; the mistake has been noticed by S.V. Konjagin).

3. The main results: Theorem I.3.8 and Corollary I.3.11 follow [142] (1975); the other ones follow [149]. The results of this section would undoubtedly loss much of their value, had we not witnessed a remarkable progress toward the solution of the existence problem (E_m). We have employed the results of papers [36], [37] and [44] only. For other facts concerning the existence of a minimal projection consult e.g. [48], [49], [114],[115],[133].

4. This section has been inspired by the papers of G. Godini [73]-[75] (1980-81) and also by the well known example due to J.Lindenstrauss [120] (1964). The main results were published partially in 1982 [150] (e.g. Proposition I.4.4 and Example I.4.18) and partially announced in 1984 [152]. Detailed proofs first appeared in 1985 in [156].

Chapter II

Minimal projections onto codimension one subspaces and a related mathematical programming problem

§ 1. Preliminaries and supplementary notations

Throughout this chapter, characters x, y, x^0, x^1,... are used to denote points of an n-dimensional real vector space $\mathbb{R}^n$; lower indices indicate coordinates, i.e., we write $x = (x_1,\ldots,x_n) \in \mathbb{R}^n$ (short $x = (x_i)$). Symbol 0 stands both for the zero element of $\mathbb{R}^n$ and for number zero. For $f \in \mathbb{R}^n$, $J_f = \{i \in \Gamma(n): f_i \neq 0\}$.

For $x \in \mathbb{R}^n$, we write $x > 0$ (resp. $x \geq 0$) iff $x_i > 0$ (resp. $x_i \geq 0$) for all $i \in \Gamma(n)$. $\mathbb{R}_+$ is the set $\{x \in \mathbb{R}: x \geq 0\}$.

Except the sections 6,7,8, we consider Banach spaces $B = (\mathbb{R}^n, \|\cdot\|)$. The dual space B^* can be also identified with $\mathbb{R}^n$, now equipped with the dual norm to the original one; i.e., given $f \in B^*$ we have

$$\|f\| = \sup\{|f(x)| \, | x \in S_B\} = \sup\{\sum_{i=1} f_i \cdot x_i : \|x\| = 1\}.$$

We denote by B_+ the nonnegative cone $\{x \in B: x \geq 0\}$.

With each $f \in B^* \setminus \{0\}$ we associate the family of operators

$$P_{f,z} = I - f \otimes z : B \to B \qquad (2.1.1)$$

indexed by elements $z \in B$ and defined by

$$P_{f,z}(x) = x - f(x) \cdot z \text{ for } x \in B.$$

When f is fixed we write P_z for $P_{f,z}$. If $z \in f^{-1}(1)$, P_z is a projection onto $D = f^{-1}(1)$. Thus we have (see [21]) a one-to-one correspondence beetween points of $f^{-1}(1)$ and projections of B onto D. Denote

$$q(f) = \inf \{\|P_{f,z}\|: z \in f^{-1}(1)\} =$$

$$= \inf \{\|P\|: P: B \to f^{-1}(0) \text{ a projection}\}, \qquad (2.1.2)$$

$$\mathcal{G}_f = \{z \in f^{-1}(1): \|P_{f,z}\| = q(f)\} \qquad (2.1.3)$$

Thus $\mathcal{G}_f$ is the set of points of $f^{-1}(1)$ which correspond to minimal projections.

The uniqueness of a minimal projection onto $f^{-1}(0)$ is expressed by the relation

$$\text{card } \mathcal{G}_f = 1.$$

The standard compactness argument shows that (in the finite-dimensional case) a minimal projection always does exist. It is also well known that a linear operator defined on a finite-dimensional Banach space attains its norm at an extremal point of the unit ball. Hence,

$$\|P_{f,z}\| = \sup\{\|x - f(x)\cdot z\| : x \in X\} \text{ if ext } B \subset X \subset S_B \qquad (2.1.4)$$

(Here ext B stands for ext W_B).

A set $Y \subset S_{B^*}$ is said to be **norming** if and only if

$\|x\| = \sup\{g(x) : g \in Y\}$ for any $x \in B$.

Equality (2.1.4) yields

Proposition II.1.1. If $Y \subset S_{B^*}$ is norming and if ext $B \subset X \subset S_B$, then

$$\|P_{f,z}\| = \sup\{g(x) - f(x)\cdot g(z) : x \in X, g \in Y\}. \qquad (2.1.5)$$

Observe t at every codimension 1 subspace of B is the kernel of a linear functional of norm 1. Therefore we may without damage confine attention to operators $P_{f,z}$ for $f \in S_{B^*}$, excluding the trivial case of dim B = 1.

Let $n \in \mathbb{N}$. Denote by E(n) the set of all n-tuples $\varepsilon = (\varepsilon_1,\ldots,\varepsilon_n)$ (short: $\varepsilon = (\varepsilon_i)$) with $\varepsilon_i \in \{-1,1\}$, and by $\Pi(n)$ the set of all permutations $\sigma : \Gamma(n) \to \Gamma(n)$. The elements of E(n) and of $\Pi(n)$ will be identified with the automorphisms of $\mathbb{R}^n$ defined for $x = (x_i) \in \mathbb{R}^n$ by

$$\varepsilon(x) = (\varepsilon_i \cdot x_i),\ \sigma(x) = (x_{\sigma(i)});\ (\varepsilon\circ\sigma)(x) = (\varepsilon_i \cdot x_{\sigma(i)}). \qquad (2.1.6)$$

Clearly, if $B = (\mathbb{R}^n, \|\cdot\|)$, $x \in B$, $f \in B^*$, then

$$(\varepsilon\circ\sigma)(f)((\varepsilon\circ\sigma)(x)) = \sum_{i=1}^{n} \varepsilon_i \cdot f_{\sigma(i)} \cdot \varepsilon_i \cdot x_{\sigma(i)} = \sum_{j=1}^{n} f_j \cdot x_j = f(x). \qquad (2.1.7)$$

Definition II.1.2. A Banach space $(\mathbb{R}^n, \|\cdot\|)$ is called **symmetric** (resp. **reflection invariant**) if $\|(\varepsilon\circ\sigma)(x)\| = \|x\|$ (resp. $\|\varepsilon(x)\| = \|x\|$) for all $x \in B$, $\sigma \in \Pi(n)$, $\varepsilon \in E(n)$. (See [155]).

Example II.1.3. Let $B = (E^n \oplus [e_{n+1}])_\infty$, where $e_{n+1} = (0,\ldots,0,1) \in \mathbb{R}^{n+1}$ ($n \geq 2$). B is easily seen to be reflection invariant but not symmetric.

Example II.1.4. Suppose that the unit ball of B is a zonotope, i.e. a polyhedron equal to the vector sum of a finite number of segments in $\mathbb{R}^n$; these polyhedra ar characterized by the property of having all two-dimensional faces centrally symmetric. In virtue of the result of P.Mc. Mullen [134], B is then a reflection invariant space. (See [32],p.180).

Remark II.1.5. If B is either symmetric or reflection invariant, then so its dual (as it not hard to check).

Proposition II.1.6. Suppose B is a symmetric space (resp. a reflection invariant space). Let $s = \varepsilon\circ\sigma$ (resp. $s = \varepsilon$) be an authomorphism of B ($\varepsilon \in E(n)$, $\sigma \in \Pi(n)$). Then

$$\|P_{f,z}\| = \|P_{sf,sz}\|, \tag{2.1.8}$$

$$\mathcal{G}_{sf} = s(\mathcal{G}_f) \tag{2.1.9}$$

hold for any $z \in B$, $f \in B^*\setminus\{0\}$.

Proof. Let B, s, ε, σ, z, f be as stated. It follows directly from Definition 1.2 that s(ext B) = ext B. Hence, in view of (2.1.4) and (2.1.7), $\|P_{f,z}\| = \sup\{\|x - f(x)\cdot z\|: x \in \text{ext } B\} = \sup\{\|s(x - f(x)\cdot z\|: x \in \text{ext } B\} =$
$= \sup\{\|sx - sf(sx)\cdot sz\|: x \in \text{ext } B\} = \sup\{\|x - sf(x)\cdot sz\|: x \in \text{ext } B\} =$
$= \|P_{sf,sz}\|$.

For the proof of (2.1.9) we first verify that q(f) = q(sf). Let $z\in\mathcal{G}_f$. Then by (2.1.7) $sz \in (sf)^{-1}(1)$ and so, by (2.1.8), $q(sf) \geq \|P_{sf,sz}\| =$
$= \|P_{f,z}\| = q(f)$. On the other hand, if $z \in \mathcal{G}_{sf}$ then $z_1 = s^{-1}z \in f^{-1}(1)$ and hence, $q(f) \geq \|P_{f,z_1}\| = \|P_{sf,sz_1}\| = q(sf)$. Thus q(f) = q(sf) and also $\mathcal{G}_{sf} = s(\mathcal{G}_f)$. ***

Consequently, when considering minimal projections onto codimension 1 subspaces of reflection invariant spaces it is sufficient to restrict attention to functionals f and projections $P_{f,z}$ with $f \in S_{B^*}$, $f \geq 0$; in the case of a symmetric space it may be additionally assumed that

$$f_1 \geq f_2 \geq \ldots \geq 0. \tag{2.1.10}$$

§ 2. A mathematical programming problem related to minimal projections

By a mathematical programming problem we mean in this paper a pair [M,L], where M is a nonempty compact subset of $\mathbb{R}^n$ and $L: M \to \mathbb{R}$ is the restriction of a linear functional to M. Denote

$M^{\perp} = \{x \in M: L(x) = \sup L(M)\}$

and call the points of $M^{\perp}$ optimal solutions of the problem. Elements of M are called feasible solutions; L is the objective function. We say that a problem [M,L] is of linear type in broad sense when M is a convex set, and in restricted sense when M is the intersection of a finite number of half-spaces with the non-negative orthant $\mathbb{R}^+_n$.

Definition II.2.1. Let $B = (\mathbb{R}^n, \|\cdot\|)$, $f \in B^* \setminus \{0\}$, $f \geq 0$. By the Bf-problem we mean the mathematical programming problem $[M,L]$ with

$$M = Q \cap {}^{\perp}K, \text{ where } Q = \{z \in B: \|P_{f,z}\| \leq q(f)\},$$

$$^{\perp}K = \{z \in B_+: \sup\{f_i \cdot z_i : 1 \leq i \leq n\} \leq 1\}; \qquad (2.2.1)$$

$$L = f|_M \qquad (2.2.2)$$

Proposition II.2.2. With $B = (\mathbb{R}^n, \|\cdot\|)$, $f \in B^* \setminus \{0\}$, $f \geq 0$, Bf-problem is of linear type in broad sense.

Proof. We have to verify that Q is a convex bounded set. That Q (hence also M) is bounded, follows directly from the definition of Q and the fact that $\lim_{\|z\| \to \infty} \|P_{f,z}\| = \infty$.

Take $z^1, z^2 \in Q$, $z^1 \neq z^2$, and λ_1, λ_2 arbitrary positive numbers with $\lambda_1 + \lambda_2 = 1$. Let $z = \lambda \cdot z^1 + \lambda \cdot z^2$. Then $f(z) = \lambda_1 \cdot f(z_1) + \lambda_2 \cdot f(z^2) = 1$, so that $z \in f^{-1}(1)$. According to (2.1.1),

$\|P_{f,z}\| \leq \sup\{\|x - f(x) \cdot z\| : \|x\| \leq 1\} \leq \lambda_1 \cdot \sup\{\|x - f(x) \cdot z_1\| : \|x\| \leq 1\} +$

$+ \lambda_2 \cdot \sup\{\|x - f(x) \cdot z^2\|; \|x\| \leq 1\} \leq \lambda_1 \cdot q(f) + \lambda_2 \cdot q(f) = q(f)$, i.e. $z \in Q$.

Since λ_1, λ_2 were chosen arbitrarily, the whole segment $[z^1, z^2]$ is contained in Q. ***

Corollary II.2.3. Let $B = (\mathbb{R}^n, \|\cdot\|)$, $f \in B^* \setminus \{0\}$, $f \geq 0$, and let $[M,L]$ be the corresponding Bf-problem. Then

$$M^{\perp} \supset \{z \in \text{ext } M : L(z) = \sup L(\text{ext } M)\}.$$

Proof. This is a direct consequence of Proposition 2.2 and the fact that a linear functional on a convex compact set M attains its upper bound at a point of ext M. (See e.g. [53]). ***

Remark II.2.4. Suppose the unit ball of B is a polyhedral set (i.e. ext B is finite) and consequently S_{B^*} contains a finite norming set. Then, by Proposition 1.1, Q is the intersection of finitely many half-spaces.

Theorem II.2.5. Let $B = (\mathbb{R}^n, \|\cdot\|)$, $n \geq 2$, $f \in B^* \setminus \{0\}$, $f \geq 0$, $q(f) > 1$ and consider the associated Bf-problem $[M,L]$. Suppose that

$$\|P_{tz}\| < \|P_z\| \text{ for all } z \in M, \ 0 \leq t < 1; \qquad (2.2.3)$$

$$\mathcal{G}_f \cap B_+ \neq \emptyset. \qquad (2.2.4)$$

Then $\mathcal{G}_f \cap B_+ = M^{\perp}$; hence, in particular, card $\mathcal{G}_f \geq$ card $M^{\perp}$.

Proof. Select $z \in \mathcal{G}_f \cap B_+$; i.e. $f(z) = 1$, $\|p_z\| = q(f)$, $z \geq 0$. Since

$f \geq 0$, we get $f_i \cdot z_i \leq \sum_{j=1}^{n} f_j \cdot z_j = 1$ for each $i \in \Gamma(n)$, and so $z \in {}^{\perp}K$. Then

$$\mathscr{G}_f \cap B_+ = \mathscr{G}_f \cap {}^{\perp}K = f^{-1}(1) \cap M \neq \emptyset. \qquad (2.2.5)$$

Consequently, $\sup f(M) \geq 1$. Assuming, there exists $y \in M$ with $f(y) = 1/t > 1$, we get $t \cdot y \in f^{-1}(1)$ and by (2.2.3) $\|P_{ty}\| < q(f)$, in contradiction to the definition of $q(f)$ as the minimal norm of a projection. Hence,

$$M^{\perp} = f^{-1}(1) \cap M = \mathscr{G} \cap B_+. \text{***}$$

Notice that condition $q(f) > 1$ has been directly employed in the proof of the theorem. Formally, the theorem remains true without this assumption, because in that case condition (2.2.3) can in no way be fulfilled; the statement is false, as well ($0 \in M^{\perp}$, but $0 \notin f^{-1}(1) \supset \mathscr{G}_f$).

———————— * ————————

To be able to apply Bf-problem we must know how to compute the norms of occurring operators $P_{f,z}$.

Assume that $B = (\mathbb{R}^n, \|\cdot\|)$ is a symmetric space. For any $x \in B$, $T \subset B$ define

$$\Delta(x) = \{(\varepsilon \circ \sigma)(x) : \varepsilon \in E(n),\ \sigma \in \Pi(n)\}, \qquad (2.2.6)$$
$$\Delta(T) = \bigcup_{x \in T} \Delta(x).$$

Remark II.2.6. Fix $x \in B$. Consider the equivalence relation on $\Pi(n)$: $\sigma \simeq u$ iff $\sigma x = ux$. Let $\Lambda(x) \subset \Pi(n)$ be the set of permutations containing exactly one representative of each equivalence class. Then

$$\Delta(x) = \{(\varepsilon \circ \sigma)(x) : \varepsilon \in E(n),\ \sigma \in \Lambda(x)\}. \qquad (2.2.7)$$

Example II.2.7. a) $B = l_\infty^n$, $e = (1,\dots,1) \in B$. Then $\text{ext } B = \Delta(e) = E(n)$, viewed as a subset of l_∞^n.

b) $B = l_1^n$, $e^1 = 1,0,\dots,0) \in B$. Then $\text{ext } B = \Delta(e^1)$.

c) $B = l_p^n$, $1<p<\infty$. Then $\text{ext } B = \Delta(S_B \cap B_+)$.

For any vector $y = (y_i) \in B$ we write $|y| = (|y_i|)$.

Proposition II.2.8. Let B be a symmetric space, $\dim B = n \geq 2$, $f \in B^* \setminus \{0\}$, $f \geq 0$, $z \in B$. Let $\{e_i\}_{i \in \Gamma(n)}$ be the canonical basis of $\mathbb{R}^n$. Assume, $T \subset B$, $\Delta(T) = \text{ext } B$. Then

$$\|P_{f,z}\| = \max\{\|\sum_{i=1}^{n} |(\varepsilon_i \cdot x_{\sigma(i)} - z_i \cdot (\sum_{j=1}^{n} \varepsilon_j \cdot f_j \cdot x_{\sigma(j)}| \cdot e^i\|\} \qquad (2.2.8)$$

or, which is the same,

$$\|P_{f,z}\| = \max\{\|\sum_{i=1}^{n} |(\varepsilon_i \cdot x_{\sigma(i)} \cdot (1 - f_i \cdot z_i) - z_i \cdot \sum_{j \neq i} \varepsilon_j \cdot f_j \cdot x_{\sigma(j)})| \cdot e^i\|\} \qquad (2.2.9)$$

maximum being taken over all $x \in T$, $\sigma \in \Lambda(x)$, $\varepsilon \in E(n)$.

Proof. Follows immediately from Definition 1.2 and formulas (2.1.4), (2.2.6), (2.2.7).***

§ 3. Space $B = l^n_\infty$. Bf - problem

Proposition II.3.1. Let $B = l^n_\infty$, $n \geq 2$, $f \in S_B*$, $f \geq 0$. Then

$$\|P_{f,z}\| = \max(|1 - f_i \cdot z_i| + |z_i| \cdot (1 - f_i) : i \in \Gamma(n)). \tag{2.3.1}$$

Proof. For $\varepsilon \in E(n)$ write

$$r_i(\varepsilon) = \varepsilon_i \cdot (1 - f_i \cdot z_i) - z_i \cdot \sum_{\substack{j=1 \\ j \neq i}}^{n} \varepsilon_j \cdot f_j .$$

Clearly,

$$\max(|r_i(\varepsilon)|; \varepsilon \in E(n)) = \varepsilon_i \cdot (1 - f_i \cdot z_i) + |z_i| \cdot \sum_{\substack{j=1 \\ j \neq i}}^{n} |f_j| =$$

$$= |1 - f_i \cdot z_i| + |z_i| \cdot (1 - f_i);$$

this follows from the conditions $f_i \geq 0$, $\sum_{i=1}^{n} f_i = \|f\| = 1$, $i \in \Gamma(n)$. Now, the statements of Example 2.7 together with Proposition 2.2 result in

$$\|P_{f,z}\| = \max \|r_i(\varepsilon)\| = \max(\max(|r_i(\varepsilon)| : i \in \Gamma(n)) : \varepsilon \in E(n)) =$$

$$= \max(\max(|r_i(\varepsilon)| : \varepsilon \in \Gamma(n)) : i \in \Gamma(n)).***$$

For the sequel we prepare a lemma, which is a direct corollary of the last proposition.

Lemma II.3.2. Let $B = l^n_\infty$, $n \geq 3$, and let $f \in S_B*$ satisfy $0 \leq f_i \leq 1/2$ for all $i \in \Gamma(n)$. Then

$$\|P_{f,z}\| = \max(V_i(z_i) : i \in \Gamma(n)), \tag{2.3.2}$$

where $V_i(t) = |1 - f_i \cdot t| + (1 - f_i) \cdot |t|$, $t \in \mathbb{R}$.

Functions V_i are decreasing for $t < 0$, nondecreasing for $t > 0$, and $V_i(0) = 1$.

Proof. It suffices to apply Proposition 2.8 observing that each V_i is continuous and piecewise differentiable, $V_i(t) = 1 - t$ for $t \leq 0$ and $V_i'(t) \in \langle 1, 1-2 \cdot f_i \rangle$ for $t > 0$.***

The following statement is an n-dimensional analogue of a result of [21] concerning the infinite dimensional space c_o (See also [45]).

Corollary II.3.3. Let $B = l_\infty^n$, $n \geq 2$, $f \in S_{B^*}$, $f \geq 0$. Then $q(f) > 1$ iff $\max f_i < 1/2$ (which is possible only if $n \geq 3$).

Proof. a) If $f_k \geq 1/2$ for certain $k \in \Gamma(n)$, we take $z_k = (1/z_k)\cdot e_k \in f^{-1}(1)$ and obtain $\|P_{f,k}\| = 1$, so that $q(f) = 1$.

b) Let z be the element of $f^{-1}(1)$ corresponding to the minimal projection P_z. Take $k \in \Gamma(n)$ so that $z_k \neq 0$. Then by Lemma 3.2 $q(f) = \|P_z\| \geq V_k(z_k) > V_k(0) = 1$.***

Corollary II.3.4. Let $B = l_\infty^n$, $n \geq 3$, and let $f \in S_{B^*}$ satisfy $f \geq 0$, $q(f) = 1 + u > 1$. Consider the Bf - problem $[M, f|_M]$. Then

$$M = \{z \in B_+ : z_i \leq u/(1-2\cdot f_i), i \in \Gamma(n)\}. \tag{2.3.3}$$

Proof. Let $V = \{z \in B_+ : z_i \leq u/(1-2\cdot f_i), i \in \Gamma(n)\}$ and let K, Q be the sets defined in (2.2.1). By Corollary 3.3, $1-2\cdot f_i > 0$. Hence,

$$M = \{x \in B_+ : 1 + \max (1-2\cdot f_i)\cdot z_i \leq q(f)\}.$$

For $z \in {}^\perp K$ we have

$$|1 - f_i\cdot z_i| + |z_i|(1 - f_i) = 1 + (1 - 2\cdot f_i)\cdot z_i \quad (i \in \Gamma(n)).$$

Thus $V \cap {}^\perp K = Q \cap {}^\perp K$. To end the proof it remains to show that $V \subset {}^\perp K$.

For each $j \in J_f$ let $z_j = (z_i^j) \in f^{-1}(1)$, where $z_i^j = 0$ for $i \neq j$, $z_j^j = 1/f_j$. Then $u = q(f) - 1 \leq \|P_{z^j}\| - 1 = (1 - 2\cdot f_j)/f_j$, and so $u/(1-2\cdot f_j) \leq 1/f_j$. This holds for all $j \in J_j$. Hence, if $y \in V$ then $0 \leq y_j \leq 1/f_j$ for all $j \in J_f$, showing that $V \subset {}^\perp K$.***

Lemma II.3.5. Let $B = l_\infty^n$, $n \geq 3$, and let $f \in S_{B^*}$ satisfy $f \geq 0$, $q(f) > 1$. For any $z \in B$ let $\dot{z} = (\dot{z}_i)$, where $\dot{z}_i = z_i$ for $i \in J_f$ and $\dot{z}_i = 0$ for $i \in \Gamma(n) \setminus J_f$. Then

$$\|P_{tz}\| < \|P_z\| \text{ for all } z \in B \setminus \{0\},\ 0 \leq t < 1; \tag{2.3.4}$$

$$\|P_{\dot{z}}\| < \|P_z\| \text{ for all } z \in B; \tag{2.3.5}$$

$$z \in \mathcal{G}_f \text{ implies } \dot{z} \in \mathcal{G}_f \cap B_+; \tag{2.3.6}$$

$$\mathcal{G}_f \cap B_+ \neq \emptyset;\ \mathcal{G}_f = \mathcal{G}_f \cap B_+ \text{ if } f > 0. \tag{2.3.7}$$

Proof. (2.3.4) and (2.3.5) follow immediately from Lemma 3.2. Statemants (2.3.7) are obviously forced by (3.2.6). It remains to prove (2.3.6).

Fix an element z in $\mathcal{G}_f$. Evidently, $f(\dot{z}) = f(z)$, whence $\dot{z} \in \mathcal{G}_f$, in view of (2.3.5). Assume $\dot{z} \notin B_+$; then $f_k\cdot z_k < 0$ for a certain $k \in J_f$. Now, look at the vector $y = (y_i)$, $y_i = z_i$ for $i \neq k$, $y_k = 0$. Writing $1/t = f(y)$ we obtain $1/t = f(y) = f(z) + |f_k \cdot z_k| > f(z) = 1$, i.e. $0 < t < 1$.

By virtue of Lemma 3.2, $\|P_y\| \le \|P_z\| = q(f)$. Further, we have $t \cdot y \in f^{-1}(y)$, whence, by (2.3.4), $\|P_{ty}\| < \|P_y\| \le q(f)$ - a contradiction, $q(f)$ denoting the minmal norm of a projection onto $f^{-1}(0)$.***

Theorem II.3.6. Let $B = l^n_\infty$, $n \ge 3$, and let $f \in S_{B^*}$ satisfy $f \ge 0$ and $\max f_i < 1/2$. Then

$$u = q(f) - 1 = (\sum_{i=1}^{n} f_i/(1-2\cdot f_i))^{-1}; \tag{2.3.8}$$

moreover, $\mathcal{G}_f \cap B_+ = M^\perp$ (the set of optimal solutions of the Bf - problem) and we have

$$\mathcal{G}_f = \left\{ z \in B: \begin{array}{l} z_i = u/(1-2\cdot f_i) \text{ for } i \in J_f \\ |z_i| \le u \text{ for } i \in \Gamma(n) \setminus J_f \end{array} \right\}. \tag{2.3.9}$$

Consequently, card $\mathcal{G}_f = 1$ iff $f > 0$.

Proof. According to Corollary 3.4, the global maximum of $f|_M$ (equal to 1, by Theorem 2.5) is attained at extremal points of M. In the case under consideration, this maximal value can be attained only at points $z \in M$ whose coordinates included in J_f are the greatest possible, i.e. with $z_i = u/(1-2\cdot f_i)$ for $i \in J_f$. Hence

$$1 = \sum_{i=1}^{n} f_i \cdot z_i = \sum_{i \in J_f} f_i \cdot z_i = \sum_{i \in J_f} f_i \cdot u/(1-2\cdot f_i) = (\sum_{i=1}^{n} f_i/(1-2\cdot z_i)).$$

Thus (2.3.8) is proved. The equality $\mathcal{G}_f \cap B_+ = M^\perp$ follows directly from Lemma 3.5 and Theorem 2.5.

According to the definition of $\mathcal{G}_f$ (formula (2.1.3)), a point z is in $\mathcal{G}_f$ iff $f(z) = 1$ and $\|P_z\| \le 1 + u$. Thus, by (2.3.6), Theorem 2.5 and Lemma 3.2, if $z \in \mathcal{G}_f$ then $z_i = u/(1-2\cdot f_i)$ for all $i \in J_f$. The condition $\|P_z\| \le 1 + u$, together with Lemma 3.2, implies the constraint: $V_i(z_i) \le 1 + u$ for all $i \in J_f$ and $|z_i| \le u$ for $i \notin J_f$ (for if $f_i = 0$ then $V_i(t) = 1 + |t|$).***

§ 4. Space $B = l^n_1$, Bf - problem

In this section we compute the norms of operators $P_{f,z}$ in l^n_1 and discuss the associated Bf - problem. As an illustration we examine an example in which the Bf - problem is applied to estabilish the uniqueness of optimal strategy in a certain matrix game; all the needed facts and notions of theory of games are given in the form of supplementary remarks after the example. The technique for solving the Bf - problem and criteria for the uniqueness of minimal projections in l_1^n will be discussed in the next section.

Since l_1^n is a symmetric space, we may freely assume that the functional f under consideration fulfils condition (2.1.10), i.e.

$$1 = f_1 \geq f_2 \geq \ldots \geq 0.$$

Recall that the norms in B and B^* are now given by $\|z\| = \sum_{i=1}^{n} |z_i|$, $\|f\| = \max |f_i|$ for $z \in B$, $f \in B^*$. Given $f \in B^*$, we let

$$^{\perp}K = \{z \in B_+ : \max f_i \cdot z_i \leq 1\}.$$

Proposition II.4.1. Let $B = l_1^n$, $n \geq 2$, and let $f \in S_{B^*}$ satisfy $f \geq 0$. Write $r_i(z) = (\|z\| - 2\cdot z_i)\cdot f_i$, $z \in B$, $i \in \Gamma(n)$. Then

$$\|P_z\| = \max\{|1-f_i \cdot z_i| + (\|z\|-2\cdot z_i)\cdot f_i : i \in \Gamma(n)\} \tag{2.4.1}$$

and

$$\|P_z\| = 1 + \max\{r_i(z) : i \in \Gamma(n)\}, \text{ where } z \in {}^{\perp}K. \tag{2.4.2}$$

Proof. It suffices to apply Example 2.7.b) and Proposition 2.8.***

Lemma II.4.2. Let $B = l_1^n$, $n \geq 3$, and let $f \in S_{B^*}$ satisfy $f \geq 0$ and $q(f) = 1 + u. > 1$. Then $\mathcal{G}_f \subset B_+$.

Proof. Assume that $z \in f^{-1}(1)$ and z has at least one negative coordinate. We have to show that $z \notin \mathcal{G}_f$, i.e., that there exists $y \in f^{-1}(1)$ with $\|P_y\| < \|P_z\|$. Let $I_z = \{i : z_i \geq 0\}$. Then $\sum_{i \in I_z} f_i \cdot z_i \geq \sum_{i=1}^{n} f_i \cdot z_i = 1$. Hence,

$$0 < t = (\sum_{i \in I_z} f_i \cdot z_i)^{-1} \leq 1.$$

Choose $y \in B$ with $y_i = 0$ for all $i \in \Gamma(n) \setminus I_z$ and $y_i = t \cdot z_i$ for $i \in I_z$. Clearly, $y \in {}^{\perp}K$ and $y \in B_+ \cap f^{-1}(1)$. To get the desired inequality $\|P_y\| < \|P_z\|$, it is enough to show that $r_i(y) + 1 < \|P_z\|$, $i \in \Gamma(n)$. We distinguish two cases.

a) $t = 1$. Then

$$f(z) = \sum_{i=1}^{n} f_i \cdot z_i = 1 = \sum_{i \in I_z} f_i \cdot z_i$$

and since $f \geq 0$, we get $f_i = 0$ when $z_i < 0$. Hence, for each i with $z_i<0$ we have $1 + r_i(y) = 1 < q(f) \leq \|P_z\|$. The definition of y clearly implies $\|y\| < \|z\|$. Thus, for i with $z_i = 0$ we get $1 + r_i(y) = 1 + \|y\|\cdot f_i <$ $< 1 + \|z\|\cdot f_i \leq \|P_z\|$. Finally, if $z_i > 0$ and $f_i = 0$, then $1 + r_i(y) <$ $< \|P_z\|$, and if $z_i > 0$ and $f_i > 0$, then $y_i = z_i$ and $1 + r_i(y) =$ $= 1 + (\|y\| - 2\cdot z_i)\cdot f_i < 1 + (\|z\| - 2\cdot z_i)\cdot f_i \leq (1-f_i\cdot z_i) + (\|z\|-z_i)\cdot f_i \leq$ $\leq |1-f_i\cdot z_i| + (\|z\|-z_i)\cdot f_i \leq q(f) \leq \|P_z\|$.

b) $t < 1$.

Then obviously $t\cdot\|z\| \geq \|y\|$. If i is such that $2\cdot z_i \geq \|z\|$ (and hence $z_i\geq 0$, $y_i = t\cdot z_i$), then $1 + r_i(y) = 1 + (\|y\|-2\cdot y_i)\cdot f_i \leq 1 + t\cdot(\|z\|-2\cdot z_i)\cdot f_i \leq 1$ $< q(f) \leq \|P_z\|$. And if $2\cdot z_i < \|z\|$, then $1 + r_i(y) \leq 1 + t\cdot(\|z\|-2\cdot z_i)\cdot f_i <$ $< 1 + (\|z\|-2\cdot z_i)\cdot f_i < \|P_z\|$.***

Theorem II.4.3. Let $B = l_1^n$, $n > 2$, and let $f \in S_{B^*}$ satisfy $f \geq 0$, $q(f) =$ $= 1 + u > 1$. Consider the Bf - problem $[M, f|_M]$. For each $i \in J_f$ let Q_i denote the half-space in B defined by the inequality

$$\sum_{j=1}^{n} z_j - 2\cdot z_i \leq u/f_i. \qquad (2.4.3)$$

For $i \in \Gamma(n) \setminus J_f$ let $Q_i = Q_o = \{z \in B: f(z) \geq 1\}$. Further, let $Q_{n+i} = \{z \in B: z_i \geq 0\}$ for all $z \in \Gamma(n)$.

Finally, let ∂Q_j be the boundary of Q_j $(j \in \Gamma(2n))$.

Then

$$M = \bigcap_{i\in J_f} Q_i \cap \bigcap_{j=n+1}^{2n} Q_j \qquad (2.4.4)$$

and

$$\mathcal{G}_f = M^{\perp} = M \cap Q_o = M \cap \partial Q_o \qquad (2.4.5)$$

(This notation is introduced for further use in section 5). Clearly $\bigcap_{i=n+1}^{2n} Q_i = B_+$, $\partial Q_o = f^{-1}(1)$.

Proof. Write $W = \{z \in B_+: \|z\| - 2\cdot z_i \leq u/f_i$ for $i \in J_f\}$. Choose arbitrarily an element $z \in W$. Assuming that $1/t = f(z) > 1$, we get $t\cdot z \in f^{-1}(1)$, whence by (2.4.2), $\|P_{tz}\| \leq 1 + t\cdot u < 1 + u = q(f)$, in contradiction to the definition of $q(f)$. Consequently $f(z) \leq 1$, and so $z \in {}^{\perp}K$. Hence, $W \subset {}^{\perp}K$, and we obtain in view of (2.2.1) and (2.4.2) $W = W \cap {}^{\perp}K = M$, i.e. formula (2.2.4).

If $z \in Q \cap {}^{\perp}K \setminus \{0\}$ and $0 \le t < 1$, then by (2.4.2) $t \cdot z \in {}^{\perp}K$ and $\|P_{tz}\| = 1 + \max\{r_i(t \cdot z) : i \in \Gamma(n)\} = 1 + t \cdot \max r_i(z) < 1 + \max r_i(z) = \|P_z\|$. This, together with Lemma 4.2 and Theorem 2.5, gives (2.4.5).***

Corollary II.4.4. Let $B = l_1^n$, $n \ge 3$, and let $f \in S_{B^*}$ satisfy $f \ge 0$, $g(f) = 1 + u > 1$. Let

$$T = \{z \in B : (\sum_{j=1}^{n} z_j - 2 \cdot z_i) \cdot f_i \le 1 \text{ for all } i \in \Gamma(n)\}.$$

Then

$$\mathcal{G}_f = u \cdot T^{\perp},$$

where $T^{\perp}$ denotes the set of optimal solutions for the problem $[T, f|_T]$.

Proof. The functions r_i occurring in Proposition 4.1 are positive-homogenous. Therefore,

$M = \{z \in B_+ : \max\{r_i(z) : i \in \Gamma(n)\} \le u\} = u \cdot \{z \in B_+ : \max r_i(z) \le 1\} = u \cdot T$

and by Theorem 4.3, $M^{\perp} = \mathcal{G}_f = u \cdot T^{\perp}$.***

Corollary II.4.5. Let $B = l_1^n, B_1 = l_1^s$, $3 \le s < n$, and let $f \in S_{B^*}$ satisfy $1 = f_1 \ge f_2 \ge \ldots \ge f_n \ge 0$, $f_s > 0$, $f_{s+1} = 0$; let $f^s = (f_1, \ldots, f_s) \in B_1^*$. Then $z \in \mathcal{G}_f$ if and only if $z_i = 0$ for $i > s$ and $z^s = (z_1, \ldots, z_s) \in \mathcal{G}_f$. Moreover, $q(f) = q(f^s)$. (Notation $q(f), \mathcal{G}_f$, and also $P_{f,y}$, see below, refer to space B).

Proof. Let $z \in \mathcal{G}_f$. By Theorem 4.3, $z \ge 0$. Write $\dot{z} = (z_1, \ldots, z_s, 0, \ldots, 0) \in B$. Of course, $\dot{z} \in f^{-1}(1)$. Also, $\|P_z\| = \|P_{\dot{z}}\|$, for otherwise $r_i(\dot{z}) < r_i(z)$ for $i \le s$ and $r_j(\dot{z}) = r_j(z) = 0$ for $j > s$, so that $\|P_{\dot{z}}\| < \|P_z\| = q(f)$, in contradiction to the minimality of $q(f)$.

According to the Proposition 4.1, if $y^s = (y_1, \ldots, y_s) \in B_1$, $y = (y_1, \ldots, y_s, 0, \ldots, 0) \in B$, then a) $\|P_{f,y}\| = \|P_{f^s,y^s}\|$, and b) $y \in f^{-1}(1)$ iff $y^s \in (f^s)^{-1}(1)$. This ends the proof.***

*

We now present an example which exhibits the interrelation between the theory of matrix games and the problem of uniqueness of minimal projections.

Example II.4.6. Let $\mathfrak{L} = (a_{ij})_{i,j \le n}$ be the square matrix with $a_{ij} = 1$ for $i \ne j$ and $a_{ij} = -1$ for $i = j$ ($i, j \in \Gamma(n)$, $n \ge 3$). The matrix game defined by $\mathfrak{L}$ has value $v \ne 0$. Consequently, by a theorem of Dantzig (see [53]) the set $Y^{\perp}$ of optimal strategies for the second player is equal to

$$Y^{\perp} = v\cdot T^{\perp},$$

$T^{\perp}$ denoting the set of optimal solutions to the programming problem

$$[T,L] \tag{2.4.6}$$

$$T = \{x \in \mathbb{R}^n : \sum_{j=1}^{n} x_j - 2\cdot x_i \le 1,\ x_i \ge 0 \text{ for all } i \le n\},$$

$$L(x) = \sum_{i=1}^{n} x_i \text{ for } x \in T.$$

Now, (2.4.6) is exactly the problem described by Corollary 4.4 for the specific functional f defined as an extention of L to all $\mathbb{R}^n$. Hence, $Y^{\perp} = u^{-1} \vee \mathcal{G}_f$.

We now solve (2.4.6). Using the chain of inequalities which define the set T we obtain $L(x) \le n/(n-2)$ for all $x \in T$.

The maximal value of L(x) equals exactly $n/(n-2) = 1 + 2/(n-2)$ and it is attained at the point $x^o = (1/(n-2)\cdot(1,\ldots,1) \in T$. Suppose x is any optimal solution of the problem. The equality $L(x) = 1 + 2/(n+2)$ combined with the inequalities defining T show that $(1+2/(n-2))-2\cdot x_i \le 1$, and so $x_i \ge 1/(n-2)$ for all $i \le n$. But since $L(x) = n/(n-2)$, we see that necessarily $x = x^o$, i.e., x^o is the only optimal solution(hence also the minimal projection from l_1^n onto $f^{-1}(0)$ is unique). Consequently, the second player in the game defined by Ł has a unique optimal strategy. (It is worth noticing that matrix Ł has appeared quite long ago in the study of operators in l_p^n, $1<p<+\infty$ (see [181],p.88).)***

Supplementary explanation. Given a matrix $L = (a_{ij})_{i\le m, j\le n}$, consider the bilinear form $F(x,y) = \sum_{i,j} a_{ij}\cdot x_i\cdot y_j$ defined on the set $X \times Y$, where

$$X = \{x \in \mathbb{R}^n : x\ge 0,\ \sum_{i=1}^{n} x_i = 1\},\ Y = \{x \in \mathbb{R}^n : \sum_{j=1}^{n} y_j = 1\}.$$

X and Y are called the sets of strategies of the first and second player, respectively. A strategy $x_o \in X$ is called optimal if $F(x^o,y) \ge F(x,y)$ for any $x \in X$, $y \in Y$; similarly, a strategy $y_o \in Y$ is optimal if $F(x,y_o) \le F(x,y)$ for any $x \in X$, $y \in Y$.

According to the classical theorem of von Neuman, optimal strategies for both players always exist. If (x_o,y_o) is any pair of optimal strategies, then $F(x_o,y_o) = \max_x \min_y F(x,y) = \min_y \max_x F(x,y)$, this common value being called the value of the game defined by the matrix L.(See e.g. [26],[69],[70],[158] for more details).

In the sequel we will need a procedure of determining the vertices of a polyhedral set M; this procedure is based on a simplex-method. (See e.g. [31],[53],[95]).

Consider the following problem $[W,L|_W]$ of mathematical programming with 2n variables

$$L(x) = \sum_{j=1}^{n} f_j \cdot x_j = \sum_{j=1}^{2n} \hat{f}_j \cdot x_j \to \max, \qquad (2.4.7)$$

where $\hat{f}_j = f_j$ for $j \le n$ and $\hat{f}_j = 0$ for $j > n$ ($f_n \ne 0, n \ge 3$). The set W of feasible solutions is defined to be the polyhedron determined by n equalities and 2n inequalities,

$$x_1 + \ldots + x_{i-1} - x_i + x_{i+1} + \ldots + x_n + x_{n+i} = u/f_i \qquad (2.4.8)$$

for $i=1,\ldots,n$, $f=(f_1,\ldots,f_n)$, $B = 1_1^n$, $u = q(f) - 1 > 0$, and

$$x_j \ge 0 \text{ for } j=1,\ldots,2n. \qquad (2.4.9)$$

Let us remark that the problem $[W,L|_W]$ is related to the Bf - problem $[M,f|_M]$ as follows:

$$M = \hat{j}(W), \quad M^{\perp} = \hat{j}(W^{\perp}), \qquad (2.4.10)$$

where $\hat{j}$ is the canonical map: $\mathbb{R}^{2n} \to \mathbb{R}^n$, given by $\hat{j}(x_1,\ldots,x_{2n}) = (x_1,\ldots,x_n)$.

Consider the system of equations

$$(\Sigma) \quad \sum_{j=1}^{n} a_{ij} x_j = d_i \quad (i \in \Gamma(n);\ J = [j_1,\ldots,j_n].$$

We call (Σ) a basic system, with basis $J = [j_1,\ldots,j_n]$, whenever (Σ) is equivalent to (2.4.8), the free terms d_i are nonnegative, and J is a sequence of distinct indices satisfying

$$a_{ij_i} = 1 \text{ and } a_{lj_i} = 0 \text{ for } l \ne i \ (i \in \Gamma(n)). \qquad (2.4.11)$$

By the basic solution of system (Σ) corresponding to basis J we mean the point $x^o \in \mathbb{R}^{2n}$ with coordinates

$$x^o_{j_i} = d_i \ (i \in \Gamma(n));\ x^o_l = 0 \text{ for } l \in \Gamma(2n) \setminus J. \qquad (2.4.12)$$

It is known (see [53]) that the canonical map $\hat{j}$ defines a one-to-one correspondence between basic solutions of the basic system equivalent to (2.4.8) and the vertices of the polyhedron M. We say that $\hat{j}(x^o)$ is the vertex corresponding to the basic system (Σ).

(For instance, the original system (2.4.8) is a basic system if we take $J = [n+1,\ldots,2n]$ for basis: the point $(0,\ldots,0,u/f_1,\ldots,u/f_n) \in \mathbb{R}^{2n}$ is its basic solution and $(0,\ldots,0) \in \mathbb{R}^n$ is the corresponding vertex of M).

Let a basis system (Σ) be given. Assume that $l \in \Gamma(n)$, $s \in \Gamma(n) \setminus J$. We define the (l,s)-operation as the replacement of system (Σ) by the new system

$$(\Sigma_o) \quad \sum_{j=1}^{2n} a'_{ij} \cdot x_j = d'_j \quad (i \in \Gamma(n));\ J' = [j_1,\ldots,j_{l-1},s,j_{l+1},\ldots,j_n],$$

where

$$a'_{lj} = a_{lj}/a_{ls} \text{ for } j \in \Gamma(2n),\ d'_l = d_l/a_{ls}, \qquad (2.4.13)$$

$a'_{is} = 0$ for $i \in \Gamma(n) \setminus \{l\}$, and

$$a'_{ij} = a_{ij} - a'_{lj}\cdot a_{is},\; d'_i = d_i - d'_l\cdot a_{is} \text{ for } i \in \Gamma(n) \setminus \{l\}, \quad j \in \Gamma(2n) \setminus \{s\}. \qquad (2.4.14)$$

Obviously, the two systems (Σ) and (Σ_o) are equivalent; system (Σ_o) is a basic one iff the new free terms d'_i are nonnegative; the (l,s)-operation is then called feasible.

The two propositions that follow will be our fundamental tool for further reasonings; they play the key role in the foundation of the simplex-method (see [31],[53],[95]).

Proposition II.4.7. Let y^1 and y^2 be the vertices of M corresponding to basic solutions of of two basic systems (Σ_1) and (Σ_2), respectively. Then y^1 and y^2 are adjacent iff (Σ_2) can be obtained from (Σ_1) by means of one feasible operation.

Proposition II.4.8. Let [W,L] be the problem of mathematical programming in which W is defined by a basic system (Σ) with the basis $J = [j_1,\dots,j_n]$ and inequalities (2.4.9) and the objective function L is given by (2.4.7). A necessary and sufficient condition in order that the basic solution of (Σ) be optimal is:

$$\hat{f}_k \le \sum_{i=1}^{n} a_{ik}\cdot\hat{f}_{jk} \text{ for all } k \in \Gamma(2n). \qquad (2.4.15)$$

*

For the sequel we adopt the convention $1/0 = \infty$, $0\cdot\infty = 0$, and we denote:

$$a_j = \sum_{i=1}^{n} f_i,\; b_j = \sum_{i=1}^{j} f_i^{-1} \; (j=1,\dots,n),\; \beta_j = b_j/(j-2),\; j\ge 3. \qquad (2.4.16)$$

We now inspect two basic systems (for simplicity, we do not any more distinguish between a system and its matrix) equivalent to (2.4.8) whose extended matrices are respectively, (2.4.17) and (2.4.18) (we employ the block-form notation):

$$\begin{array}{c|c|c|c|c|l} Z_1 & Z_2 & Z_3 & Z_4 & Z_9 & \} k \\ \hline Z_5 & Z_6 & Z_7 & Z_8 & Z_{10} & \} n-k \\ \hline k & n-k & k & n-k & 1 & \end{array} \qquad (2.4.17)$$

with basis $J = [1,2,\dots,k,n+k+1,n+k+2,\dots,2n]$, where $3 \le k \le n$

$$Z_1 = \left\{\begin{array}{c} 1\dots0 \\ \dots\dots \\ 0\dots1 \end{array}\right. \qquad Z_2 = \left\{\begin{array}{c} t\dots t \\ \dots\dots \\ t\dots t \end{array}\right. .$$

$$Z_3 = \begin{Bmatrix} -(k-3)\cdot t/2 & t/2 & \dots & t/2 \\ t/2 & -(k-3)\cdot t/2 & & \\ \dots & \dots & \dots & \dots \\ t/2 & t/2 & \dots & -(k-3)\cdot t/2 \end{Bmatrix}$$

$$Z_4 = \begin{Bmatrix} 0 \dots 0 \\ \dots\dots \\ 0 \dots 0 \end{Bmatrix} \qquad Z_9 = \begin{Bmatrix} (u/2)\cdot(\beta_k - f_1^{-1}) \\ \dots\dots\dots \\ (u/2)\cdot(\beta_k - f_k^{-1}) \end{Bmatrix}$$

$$Z_5 = \begin{Bmatrix} 0 \dots 0 \\ \dots\dots \\ 0 \dots 0 \end{Bmatrix} \qquad Z_7 = \begin{Bmatrix} -t \dots -t \\ \dots\dots \\ -t \dots -t \end{Bmatrix} \qquad Z_8 = \begin{Bmatrix} 1 \dots 0 \\ \dots\dots \\ 0 \dots 1 \end{Bmatrix}$$

$$Z_6 = \begin{Bmatrix} -(2k+2)\cdot t & -2\cdot t & \dots & -2\cdot t \\ -2\cdot t & -(2k+2)\cdot t & \dots & -2\cdot t \\ \dots & \dots & \dots & \dots \\ -2\cdot t & -2\cdot t & \dots & -(2k+2)\cdot t \end{Bmatrix}$$

$$Z_{10} = \begin{Bmatrix} u\cdot(f_{k+1}^{-1} - \beta_k) \\ \dots\dots\dots \\ u\cdot(f_n^{-1} - \beta_k) \end{Bmatrix}, \quad t = (k-2)^{-1},$$

and

$$\begin{array}{c|c|c|c|c|l} Z'_1 & Z'_2 & Z'_3 & Z'_4 & Z'_9 & \} \; k \\ \hline Z'_5 & Z'_6 & Z'_7 & Z'_8 & Z'_{10} & \} \; n-k \\ \hline k & n-k & k & n-k & 1 & \end{array} \qquad (2.4.18)$$

with basis $J = [1,2,\dots,k-1,n+1,n+k+1,\dots,2n]$, where $3 \le k \le n$,

$$Z'_1 = \begin{Bmatrix} 1 & 0 & \dots & 0 & k-3 \\ 0 & 1 & \dots & 0 & -1 \\ \dots & \dots & \dots & \dots & \dots \\ 0 & 0 & \dots & 1 & -1 \\ 0 & 0 & & 0 & 2k-4 \end{Bmatrix} \qquad Z'_2 = \begin{Bmatrix} 1/2 & \dots & 1/2 \\ 0 & \dots & 0 \\ \dots & \dots & \dots \\ 0 & \dots & 0 \\ 2 & \dots & 2 \end{Bmatrix}$$

$$Z'_3 = \begin{Bmatrix} 0 & 1/2 & 1/2 & \dots & 1/2 & 1/2 & 2-k/2 \\ 0 & -1/2 & 0 & \dots & 0 & 0 & 1/2 \\ 0 & 0 & -1/2 & \dots & 0 & 0 & 1/2 \\ \dots & \dots & \dots & \dots & \dots & \dots & \dots \\ 0 & 0 & 0 & \dots & 0 & -1/2 & 1/2 \\ 1 & 1 & 1 & \dots & 1 & 1 & 3-n \end{Bmatrix}$$

$$Z'_4 = \begin{Bmatrix} 0 & \dots & 0 \\ \dots & \dots & \dots \\ 0 & \dots & 0 \end{Bmatrix}, \quad Z'_9 = \begin{Bmatrix} (u/2)\cdot(b_k - (k-3)\cdot f_k^{-1} - 1) \\ (u/2)\cdot(f_k^{-1} - f_2^{-1}) \\ \dots\dots\dots \\ (u/2)\cdot(f_k^{-1} - f_{k-1}^{-1}) \\ u\cdot(k-2)\cdot(\beta_k - f_k^{-1}) \end{Bmatrix}$$

$$Z'_5 = \begin{Bmatrix} 0 & \dots & 0 & 2 \\ \dots & \dots & \dots & \dots \\ 0 & \dots & 0 & 2 \end{Bmatrix}, \qquad Z'_6 = \begin{Bmatrix} -2 & 0 & \dots & 0 \\ 0 & -2 & \dots & 0 \\ \dots & \dots & \dots & \dots \\ 0 & 0 & \dots & -2 \end{Bmatrix},$$

$$Z'_7 = \left\{ \begin{array}{cccc} 0 & \ldots & 0 & -1 \\ \multicolumn{4}{c}{\ldots\ldots\ldots} \\ 0 & \ldots & 0 & -1 \end{array} \right\}, \quad Z'_8 = \left\{ \begin{array}{cccc} 1 & 0 & \ldots & 0 \\ \multicolumn{4}{c}{\ldots\ldots\ldots} \\ 0 & \ldots & 0 & 1 \end{array} \right\},$$

$$Z'_{10} = \left\{ \begin{array}{c} (u/2)\cdot(f^{-1}_{k+1} - f^{-1}_k) \\ \ldots\ldots\ldots\ldots\ldots \\ (u/2)\cdot(f^{-1}_n - f^{-1}_k) \end{array} \right\}$$

To derive (2.4.17) from (2.4.8) we proceed as follows. We take a sum of first k equations of (2.4.8) and divide each side of the resulting equations by k-2, thus obtaining

$$x_1 + \ldots + x_k + k/(k-2)\cdot(x_{k+1} + \ldots + x_n) +$$

$$+ 1/(k-2)\cdot(x_{n+1} + \ldots + x_{n+k}) = n\cdot b_k/(k-2) \qquad (2.4.19)$$

Then we subtract (2.4.19) from each equation of (2.4.8) ((2.4.19) will not be included in the new system). Finally, we derive all coefficients in the initial k-1 equations by -2. As a result we obtain (2.4.17).

System (2.4.18) arises from (2.4.17) by means of the (k,n+1)-operation, which is feasible.

It goes without saying that, in order to produce basic system, we have to require that the free terms columns in (2.4.17) and (2.4.19) are nonnegative; i.e., $f^{-1}_k \le \beta_k \le f^{-1}_{k+1}$ for (2.4.17) and $\beta_k \ge f^{-1}_k$ for (2.4.18).

In the specific case of k = n, systems (2.4.17) and (2.4.18) take on the form

$$\left\{ \begin{array}{ccccccc} 1 & 0\ldots0 & 0 & -(n-3)\cdot t/2 & t/2 & \ldots t/2 \ldots & (u/2)\cdot(\beta_n - f^{-1}_1) \\ 0 & 1\ldots0 & 0 & t/2 & -(n-3)\cdot t/2 & \ldots t/2 \ldots & (u/2)\cdot(\beta_n - f^{-1}_k) \\ \multicolumn{7}{c}{\ldots\ldots\ldots\ldots\ldots\ldots\ldots\ldots\ldots\ldots} \\ 0 & 0\ldots0 & 1 & t/2 & t/2 \ldots & -(n-3)\cdot t/2 \ldots & (u/2)\cdot(\beta_n - f^{-1}_n) \end{array} \right\} \qquad (2.4.20)$$

$$\qquad\quad n \qquad\qquad\qquad\qquad\qquad\qquad n$$

with basis J = [1,...,n], $t = (k-2)^{-1}$ and

$$\left\{ \begin{array}{cccccccccc} 1 & 0\ldots0 & n-3 & 0 & 1/2 & 1/2\ldots1/2 & & 2-n/2 & (u/2)\cdot(b_n - (n-3)f^{-1}_n - 1) \\ 0 & 1\ldots0 & -1 & 0 & -1/2 & 0 \ldots & 0 & 1/2 & (u/2)\cdot(f^{-1}_n - f^{-1}_2) \\ \multicolumn{9}{c}{\ldots\ldots\ldots\ldots\ldots\ldots\ldots\ldots\ldots\ldots} \\ 0 & 0\ldots1 & -1 & 0 & 0 & 0 \ldots & -1/2 & 1/2 & (u/2)\cdot(f^{-1}_n - f^{-1}_{n-1}) \\ 0 & 0\ldots0 & 2n-4 & 1 & 1 & 1 \ldots & 1 & -(n-3) & u\cdot(n-2)(\beta_n - f^{-1}_n) \end{array} \right\} \qquad (2.4.21)$$

n time n time

with basis J = [1,...,n-1,n+1].

Theorem II.4.9. Let $B = l^n_1$, n≥3, and let $f \in S_{B^*}$ satisfy $1 = f_1 \ge f_2 \ge \ldots \ge f_n \ge 0$, $f_3 > 0$. Let

$$c_j = \min\,(f_j\cdot b_{j-1}, a_{j-1}),\ j \ge 2; \qquad (2.4.22)$$

$$k = k(f) = \max\{j : c_j - j \ge -3\}. \qquad (2.4.23)$$

Then $q(f) = 1 + u$, where

$$u = \begin{cases} 2\cdot((\beta_k - f_k^{-1})\cdot(k-2) + a_k\cdot f_k^{-1} - k)^{-1} & \text{if } a_k < k-2 \\ 2\cdot(a_k\cdot\beta_k - k)^{-1} & \text{if } a_k \geq k-2 \end{cases} \tag{2.4.24}$$

and

$$\beta_k \geq f_k^{-1}. \tag{2.4.25}$$

Proof. First of all, we observe that formula (2.4.23) is correct, since $c_s - 3 \geq -3$; this inequality follows from $f_s\cdot b_2 \geq 0$ and $a_2 > 0$. Moreover, if $a_j \geq j-2$ then $a_{j-1} \geq j-3$ because $a_{j-1} = a_j - f_j \geq j-2 - f_j \geq j-3$. Further, if $f_{j+1}\cdot b_j \geq j-2$, then $f_j\cdot b_{j-1} \geq j-3$, since $f_j\cdot b_{j-1} = f_j\,(b_j - f_j^{-1}) = f_j\cdot b_j - 1 \geq j-3$. Consequently, if $c_j - j \geq -3$ then $c_{j-1} - (j-1) \geq -3$.

We see that $k = k(f) \geq 3$. Assume that there is an index s with $3 \leq s \leq n$ such that $f_s > 0$, $f_{s+1} = 0$. Since $c_{s+1} = 0$ and $s+1 \geq 4$, we have $k(f) = k(f^s)$, where $f^s = (f_1,\dots,f_s)$. Consider the space $B^s = l_1^s$. Corollary 4.5 asserts that $q(f^s) = q(f)$, in view of the assumption $f_s > 0$, which forces $q(f) > 1$. We may thus assume, without loss of generality, that $f_n > 0$.

According to the definition of k(f), two cases are possible: (a) $a_k < k-2$, (b) $a_k \geq k-2$ (in the latter case, if $k < n$, then $f_{k+1}\cdot b_k < k-2$).

We first consider case (a): $a_k < k-2$. Then, by the definition of $k = k(f)$, $f_k\cdot\beta_k = f_k\cdot b_k\cdot(k-2)^{-1} = f_k\cdot(b_{k-1} + f_k^{-1})\cdot(k-2)^{-1} =$

$$= (f_k\cdot b_{k-1} + 1)\cdot(k-2)^{-1} \geq (k-3+1)\cdot(k-2)^{-1} = 1.$$

Hence, $f_k^{-1} \leq \beta_k$, ensuring that (2.4.18) is a basic system.

Now, the following inequalities are (jointly) necessary and sufficient for the basic solution of this system to be optimal (see Proposition 4.8, condition (2.4.15)):

$f_1 \leq f_1 \leq,\dots,\leq f_{k-1} \leq f_{k-1}$,

$f_k \leq (k-3) - a_{k-1} + 1$,

$0 \leq 1/2$ (repeated n-k times),

$0 \leq 0$, $0 \leq 1/2 - (1/2)\cdot f_2,\dots,$ $0 \leq 1/2 - (1/2)\cdot f_{k-1}$,

$0 \leq 2 - (1/2)\cdot k + (1/2)\cdot a_{k-1} - 1/2$,

$0 \leq 0$ (repeated n-k times).

Evidently, the k-th inequality is equivalent to $a_k \leq k-2$; whereas the (n+k)-th one is equivalent to $a_{k-1} \geq k-3$.

It follows from the definition of $k = k(f)$ and assumption (a) that both these inequalities are satisfied (the remaining ones are automatic), and thus the basic solution of (2.4.18) is optimal.

Let z be the vertex of M corresponding to the basic solution of (2.4.18). Theorem 4.3 implies that the objective function takes value 1 at z, and so

$$1 = f(z) = (u/2)\cdot(b_k-(k-3)\cdot f_k^{-1}-1) + (u/2)\cdot(\sum_{i=1}^{k-1} f_i\cdot(f_k-f_i^{-1})).$$

Hence,

$$u = 2\cdot(b_k-(k-3)\cdot f_k^{-1}- 1 + f_k^{-1}\cdot(a_k-f_k) + f_k^{-1}- (k-2))^{-1} =$$

$$= 2\cdot((\beta_k-f_k)\cdot(k-2) + a_k\cdot f_k^{-1}- k)^{-1} =$$

$$= 2\cdot(a_k\cdot\beta_k- k + (\beta_k -f_k^{-1})\cdot(k-2-a_k))^{-1}$$

and $q(f) = 1 + u$.

Now we consider case (b): $a_k \geq k-2$. (Here, if $k<n$ then $f_{k+1}\cdot b_k<(k-2)$. Just as in case (a) we verify that $f_k^{-1} \leq \beta_k$. If $k < n$ then $\beta_k \leq f_{k+1}^{-1}$ because $f_{k+1}\cdot\beta_k = f_{k+1}\cdot(k-2)^{-1} < 1$. Thus (2.4.17) is a basic system. Let us write down the conditions for its basic solution to be optimal:

$f_1 \leq f_1 \leq, \dots, f_k \leq f_k$,

$f_{k+1} \leq (k-2)^{-1}\cdot a_k, \dots, f_n \leq (k-2)^{-1}\cdot a_k$,

$0 \leq a_k/(2k-4) - ((k-3)/(2k-4))\cdot f_1 - f_1/(2k-4), \dots,$

$0 \leq a_k/(2k-4) - ((k-3)/(2k-4))\cdot f_k - f_k/(2k-4)$,

$0 \leq 0$ (repeated n-k times).

Since $a_k\geq k-2$ and $f_j \leq 1 (j=1,\dots,n)$, the (k+i)-th inequality is satisfied $(i=1,\dots,n)$ (the other ones being trivial).

Again, by Theorem 4.1 we infer that the objective function assumes value 1 at y, the vertex of M corresponding to the basic solution of (2.4.17). Thus

$$1 = f(y) = (u/2)\cdot(a_n\cdot\beta_n- n).$$

Consequently, $u = 2\cdot(a_n\cdot\beta_n - n)^{-1}$ and $q(f) = 1 + u$.***

Remark II.4.10. Theorem 4.9 can be also obtained by adaptation of a result of Blatter and Cheney [21] concerning the norm of a minimal projection in l_1.

5. Criterion for uniqueness of minimal projections in l_1^n

In the sequel we shall need the following theorem, which is a finite dimensional analogue of Theorem 3 in [21].

Theorem II.5.1. Let $B = l_1^n, n \geq 2$, $f \in S_{B^*}$. The equality $q(f) = 1$ holds if and only if f has no more than two nonzero coordinates (i.e. card $J_f \leq 2$); a norm 1 projection from B onto $f^{-1}(0)$ is unique iff card $J_f = 2$.

Proof. If card $J_f \leq 2$, we may assume (with no loss in generality, in view of Corollary 4.5) that $n=2$; but then both assertions of the theorem are obvious. If card $J_f \geq 3$, then by Proposition 4.7 $u > 0$, so that $q(f) > 1$.***

Now we state the main result of this section.

Theorem II.5.2. Let $B = l_1^n$, $n \geq 3$, and let $f \in S_{B^*}$ satisfy

$$1 = f_1 \geq \ldots \geq f_n \geq 0, \; f_3 > 0. \tag{2.5.1}$$

Let

$$m = m(f) = \begin{cases} k(f) & \text{if } 1/f_{k(f)} \neq \beta_{k(f)} \\ \min\{i \in \Gamma(k) \setminus \Gamma(2) : 1/f_{i+1} = \beta_{k(f)}\} & \text{otherwise.} \end{cases} \tag{2.5.2}$$

A minimal projection from B onto $f^{-1}(0)$ is unique if and only if one of the following two conditions is satisfied:

(i) $a_m > m-2$;

(ii) $a_m < m-2$, $f_2 < 1$, $a_{m-1} > m-3$.

(Symbols $k(f)$, a_m, β_i have the same meaning as in Theorem 4.9).

In particular, for n=3 (see also [45],[155]) and n=4 we obtain

Corollary II.5.3. Let $b = l_1^3$, $f \in S_{B^*}$, $f > 0$. Then card $\mathcal{G}_f = 1$.

Proof. Here m=n and $a_3 = 1 + f_2 + f_3 > 1 = 3-2$.***

Corollary II.5.4. Let $B = l_1^4$ and let $f \in S_{B^*}$ satisfy (2.5.1). Then card $\mathcal{G}_f > 1$ if and only if

$$f_2 + f_3 + f_4 = 1 \text{ and } 1 + 1/f_2 + 1/f_3 > 1/f_4$$

hold simultanously.

Proof. First observe that (2.5.1) yields $\beta_3 = 1 + 1/f_2 + 1/f_3 < \infty$, and hence the inequality of the corollary gives $f_4 > 0$.

If $m(f) = 3$, then (2.5.1) shows that $a_3 > 1$ and so, by Theorem 5.2, card $\mathscr{G}_f = 1$.

It is easy to see that $m(f) = 4$ iff $1 + 1/f_2 + 1/f_3 > 1/f_4$. In view of (2.5.1), $a_3 > 1$ and if $a_4 < 2$ then $f_2 < 1$, i.e., condition (ii) of Theorem 5.2 is fulfilled; and if $a_4 > 2$, (i) is met. Hence, in the case of $m(f) = 4$, the equality card $\mathscr{G}_f = 1$ holds iff $a_4 \neq 2$, i.e., iff the equality $f_2 + f_3 + f_4 = 1$ fails to hold. ***

Proof of Theorem 5.2.

I. Case $m(f) = n$.

Recall that M, the set of feasible solutions of the Bf-problem, is in our case a compact polyhedron. We will use the notation of Theorem 4.3 and 4.9. We now sketch the general line of the proof.

As it was shown in the proof of Theorem 4.9, when, $m(f) = n$, (hence also $k(f) = n$), the points

$z^{0} = (u/2)\cdot(\beta_n - f_1^{-1}, \ \ldots \ , \beta_n - f_n^{-1})$ and

$z^{1} = (u/2)\cdot(f_n^{-1} - f_1^{-1}, \ldots, f_n^{-1} - f_{n-1}^{-1}, f_n^{-1} - f_n^{-1}) + (u/2)\cdot(b_n - (n-2)\cdot f_n^{-1})\cdot e^{1}$

are vertices of M. (Here and before, $e^{i} = (0,\ldots,0,\underset{i}{1},0,\ldots 0) \in \mathbb{R}^{n}$, $i=1,\ldots,n$). Vertex z^{0} belongs to σQ_0 (i.e. to $f^{-1}(1)$) only if the upper line in equality (2.4.24) is the case, and z^{1} is in σQ only if the lower line is valid (in particular, both z^{0} and z^{1} are in σQ_0 when $a_n = n-2$).

According to Theorem 4.3, the hyperplane σQ_0 supports the polyhedron M. Therefore, the set $\mathscr{G}_f = M \cap \sigma Q_0$ must contain a vertex of M, call it y, which is adjacent to z (i.e., z and y are linked by an edge of M). Hence, the proof essentially reduces to finding the vertices of M adjacent to z^{0} or z^{1} and computing the values of f at these points.

Now we examine feasible operations with systems (2.4.20) and (2.4.21); vertices z^{0} and z^{1} correspond respectively to the basic solutions of these system.

We write down the vertices of M corresponding to basic solutions of the resulting systems in the form of two tables; conditions in square brackets are the restrictions imposed on the coordinates of f in order that the operation in the question be feasible (i.e. ensuring that the basic solution is nonnegative).

Table 1

Feasible operations on (2.4.20); vertices adjacent to z^0

column / row	$n+\nu,\ 2\le\nu\le n-1$	$2n$
$\nu\ne l<n$	$z^{l\nu}=(u/2)\cdot((f_l^{-1}-f_1^{-1},\dots,f_l^{-1}-f_n^{-1})+$ $+(b_n-(n-2)/f_l))\cdot e^{\nu}$; $[f_l=f_n]$	$z^{ln}=(u/2)\cdot((f_l^{-1}-f_1^{-1},\dots,f_l^{-1}-f_n^{-1})+$ $+(b_n-(n-2)/f_l))\cdot e^{n}$; $[f_l=f_{n-1};b_{n-1}-(n-3)/f_{n-1}\ge 0]$
$l=n$	$z^{\nu}=z^{n\nu}=(u/2)\cdot((f_n^{-1},f_1^{-1},\dots,f_n^{-1}-f_n^{-1})+$ $+(b_n-(n-2)/f_n)\cdot e^{\nu}$	non-feasible

Table 2

Feasible operations on (2.2.21); vertices adjacent to z^1

column / row	n	$n+\nu,$ $2\le\nu\le n-1$	$2n$
$l=1$	$w^0=z_1^1-z^1\cdot e^1+(z_1^1/(n-3))\cdot$ $(e^2+,\dots,+e^{n-1}+e^n)$, feasible if $n>3$	$w^{\nu}=z^1-z_1^1\cdot e^1+$ $+z_1^1 e^{\nu}$	non-feasible if $n>3$
$2\le l\le$ $\le n-1$	non-feasible	non-feasible	$y^l=z^1+(n-3)\cdot z_1^1\cdot e^1-$ $-z_1^1\cdot(e^1+,\dots,+e^{n-1})$; $[f_l=f_{n-1}]$
$l=n$	z^0	z^{ν} (as in Table 1)	non-feasible

The next proposition comprises all information on the vertices of M which is needed in the proof of Theorem 5.2 (in the case $m(f)=n$).

Proposition II.5.5. Assume that the functional f satisfies conditions (2.5.1) and $m(f)=n$ (i.e. $f_n>0$, $a_{n-1}\ge n-3$, $\beta_n>f_n^{-1}$). Then

a) if $a_n>n-2$ then $z^0\in\sigma Q_0$ and the adjacent vertices (see Table 1) are not in σQ_0;

b) if $a_n = n-2$ then $[z^0,z^1] \subset \partial Q_0$ and $z^1 \neq z^0$;

c) if $a_n < n-2$, $f_2 = 1$, then $[z^1,z^2] \subset \partial Q_0$ and $z^2 \neq z^1$;

d) if $a_n < n-2$, $a_{n-1} = n-3$, $f_{n-1} \neq f_n$, then $[z^1,y^{n-1}] \subset \partial Q_0$ and $y^{n-1} \neq z^1$;

e) if $a_n < n-2$, $f_2 \neq 1$ and either $a_{n-1} \neq n-3$ or $f_{n-1} \neq f_n$, then $z^1 \in \partial Q_0$ and the adjacent vertices (see Table 2) are not in ∂Q_0.

Proof. a) If $a_n > n-2$ then u is given by the lower line of formula (2.4.24), and hence

$$f(z^0) = (u/2)\cdot\sum_{j=1}^{n} f_j\cdot(\beta_j-f_j^{-1}) = (u/2)\cdot(\beta_n\cdot a_n - n) = 1.$$

If $f_l = f_n$ then

$$f(z^{l\nu}) = f(z^\nu) = (u/2)\cdot((a_n\cdot f_n^{-1}-n)+(b_n-(n-2)f_n^{-1})f_\nu) =$$

$$= (u/2)\cdot((a_n\cdot\beta_n-n)-a_n\cdot(\beta_n-f_n^{-1})+(n-2)\cdot(\beta_n-f_n^{-1})\cdot f_\nu) =$$

$$= 1 + (u/2)\cdot(\beta_n-f_n^{-1})\cdot((n-2)\cdot f_\nu-a_n) = 1 + (u/2)\cdot(\beta_n-f_n^{-1})\cdot(n-2-a_n) < 1.$$

If $f_l = f_{n-1}$ then

$$f(z^{ln}) = (u/2)\cdot(a_n\cdot f_{n-1}^{-1}-n)+(u/2)\cdot(b_n-(n-2)\cdot f_{n-1}^{-1})\cdot f_n =$$

$$= (u/2)\cdot((a_n\cdot\beta_n-n)-a_n\cdot(\beta_n-f_{n-1}^{-1})+f_n\cdot(n-2)\cdot(\beta_n-f_{n-1}^{-1})) =$$

$$= 1 - (u/2)\cdot(\beta_n-f_{n-1}^{-1})\cdot(a_n-(n-2)\cdot f_n) \le 1 - (u/2)\cdot(\beta_n-f_n^{-1})\cdot(a_n-(n-2)) < 1.$$

Before passing to proofs of (b) - (e) let us remark that if $n = 3$ then $a_3 > 1$; therefore, we may assume in these proofs that $n \ge 4$.

(b) If $a_n = n-2$, then the two expressions in formula (2.4.24) (the upper and the lower one) are equal. Thus $f(z^0) = 1$, just as in (a). Further, $f(z^1) = (u/2)\cdot a_n\cdot f_n^{-1}-n)+(u/2)\cdot(b_n-(n-2)/f_n)\cdot f_1 =$

$$= (u/2)\cdot(a_n\cdot f_n^{-1}-n+(n-2)\cdot(\beta_n-f_n^{-1})) = 1 \text{ and}$$

$$z_2^0 = (u/2)\cdot(\beta_n-f_2^{-1}) \neq (u/2)\cdot(f_n^{-1}-f_2^{-1}) = z_2^1,$$

since $\beta_n \neq f_n^{-1}$ (as $k(f) = n$). Hence, $z^0 \neq z^1$.

In the remaining cases (c) - (e) we have $a_n < n-2$ and by (2.4.12)

$$u = ((\beta_n-f_n^{-1})\cdot(n-2)+a_n\cdot f_n^{-1}-n)^{-1}.$$

The verification that $z^1 \in \partial Q_0$ is then immediate. As to the other statements:

(c) If $f_2 = f_1 = 1$ then

$$f(z^2) = (u/2)\cdot(a_n\cdot f_n^{-1}-n)+(u/2)\cdot(n-2)\cdot(\beta_n-f_n^{-1})\cdot f_2 = f(z^1) = 1$$

and

$$z_1^2 = (u/2)\cdot(f_n^{-1}-f_1^{-1}) \neq (u/2)\cdot(f_n^{-1}-f_1^{-1})+(u/2)\cdot(n-2)\cdot(\beta_n-f_n^{-1}) = z_1^1,$$

so that $z^2 \neq z^1$.

(d) Let $a_{n-1} = n-3$ (remember, $a_n < n-2$). Then

$$f(y^{n-1}) = f(z^1)+(n-3)\cdot z_{n-1}^1 - z_{n-1}^1\cdot a_{n-1} = 1 + z_{n-1}^1\cdot(n-3-a_{n-1}) = 1,$$

and evidently

$$y^{n-1} \neq z^1 \text{ iff } z_{n-1}^1 \neq 0, \tag{2.5.3}$$

which is equivalent to $f_{n-1} \neq f_n$.

(e) Let $f_2 \neq 1$ (still keeping in mind that $a_n < n-2$). Then

$$f(w^0) = f(z_1)+z_1^1\cdot(n-3)^{-1}\cdot f(e^2+ \dots +e^{n-1}+e^n) =$$

$$= 1 + z_1^1\cdot((a_n-1)/(n-3)-1) < 1, \text{ since}$$

$$z_1^1 = (u/2)\cdot(f_n^{-1}-1+(n-2)\cdot(\beta_n-f_n^{-1})) > 0.$$

Further, we have

$$f(w^\nu) = f(z^1)+z_1^1\cdot(f_\nu-f_1) \neq f(z^1) = 1 \text{ for } 2 \le \nu \le n-1;$$

$$f(z^\nu) = f(z^1+(u/2)\cdot(b_n-(n-2)/f_n)\cdot(e^\nu-e^1)) =$$

$$= 1 + (u/2)\cdot(b_n-(n-2)/f_n)\cdot(f_\nu-f_1) \neq 1 \text{ for } 2 \le \nu \le n-1.$$

Finally (see (d)),

$$f(y^l) = 1 + z_1^1\cdot(n-3-a_{n-1}) \neq 1, \text{ provided } f_{n-1}\neq f_n \text{ and } a_{n-1}\neq n-3.$$

In the case when $f_{n-1} = f_n = f_l$ we have $y^l = z^1$; see (2.5.3).***

Remark II.5.6. Proceeding along the same lines as we did in the last proof one can examine with more care all vertices listed in Tables 1 and 2 and get some more information about the set $\mathcal{G}_f$ (in the case of non-uniqueness). To obtain a complete information about that set, one has to examine vertices adjacent to those given in Tables 1 and 2, then the vertices adjacent to the new ones, and so on. The polyhedron M has finitely many vertices, and so does $\mathcal{G}_f$.

II. Reduction to the case I.

To complete the proof of Theorem 5.2, it suffices to reduce the general case to that of $m(f) = n$. This is achieved in the following proposition.

Proposition II.5.7. Let $B = 1_1^n$, $n\ge3$, and let $f \in S_B*$ satisfy conditions (2.5.1). Let $m = m(f) \ge 3$, $B_m = 1_1^m$, $f^m = (f_1,\dots,f_m) \in S_{B_m}*$. Then

$m(f^m) = m(f)$, $q(f^m) = q(f)$ and

$$\mathcal{G}_f = \{z \in B_+ : (z_1,\dots,z_m) \in \mathcal{G}_{f^m},\ z_j = 0 \text{ for } m < j \le n\}. \tag{2.5.4}$$

In particular,

card $\mathcal{G}_f$ = card $\mathcal{G}_f m$. (2.5.5)

We well need a chain of lemmas.

Lemma II.5.8. Let $B = l_1^n$, $n \geq 3$, and let $f \in S_B{*}$ satisfy (2.5.1). Then $1 < q(f) < 2$.

Proof. The inequality $1 < q(f)$ follows directly from Theorem 5.1. Let $j = \max\{i : f_i = 1\}$ and let $z \in f^{-1}(1)$ be the point with coordinates $z_i = 1/j$ for $i \leq j$ and $z_i = 0$ for $j < i \leq n$. By Proposition 4.1,

$q(f) \leq \|P_z\| = 1 + \max\{f_{j+1}, 1-2/j\} < 2$ if $j < n$ and $q(f) \leq \|P_z\| = 2-2/n$ when $j=n$. ***

Remark II.5.9. Lemma 5.8 can be proved without resorting to Proposition 4.1 and using Proposition 4.7 instead, combined with the inequality

$$(\sum_{i=1}^n \alpha_i \cdot x_i^2)\cdot(\sum_{i=1}^n 1/\alpha_i) \geq \sum_{i=1}^n x_i \qquad (\alpha_i > 0, x_i > 0,\ i \in \Gamma(n))$$

(see [13], Comments to ChapterI).

Lemma II.5.10. Let $B = l_1^n$, $n \geq 4$, $f \in S_B{*}$, $f > 0$, and let $3 \leq s < n$, $B_s = l_1^s$, $f^s = (f_1, \dots, f_s) \in B_s^*$. If $q(f) = q(f^s)$ then

$$\mathcal{G}_f \subset \{z \in B^+ : (z_1, \dots, z_s) \in \mathcal{G}_{f^s},\ z_i = 0 \text{ for } s+1 \leq i \leq n\}.$$

Proof. Choose $z \in \mathcal{G}_f$ and write $u = q(f)-1 = q(f^s) - 1 = \|P_z\| - 1$.

By Lemma 4.2, $z \geq 0$. Lemma 5.8 and Proposition 4.1 yield to the estimate

$$\|z\| - 2\cdot z_1 = (\|z\| - 2\cdot z_1)\cdot f_1 \leq \|P_z\| - 1 = u < 1,$$

which combined with the estimate $\|z\| = \|z\|\cdot\|f\| \geq f(z) = 1$ gives the inequality $z_1 > 0$. Hence,

$$0 < t = \sum_{i=1}^n f_i \cdot z_i = 1 - \sum_{i=s+1}^n f_i \cdot z_i.$$

Write $y^s = t^{-1}\cdot(z_1, \dots, z_s)$. Clearly, $y^s \in (f^s)^{-1}(1)$. By Proposition 4.1 there exists an index $i \in \Gamma(n)$ such that

$$t\cdot u \leq t\cdot(\|P_{y^s}\| - 1) = t\cdot f_i(\|y^s\| - 2\cdot y_i^s) = f_i\cdot(\|t\cdot y^s\| - 2\cdot z_i) =$$

$$= f_i\cdot(\|z\| - \sum_{j=s+1}^n z_j - 2\cdot z_i) = f_i\cdot(\|z\| - 2\cdot z_i) - f_i\cdot\sum_{j=s+1}^n z_j \leq u - f_i\cdot\sum_{j=s+1}^n z_j.$$

Consequently,

$$f_i\cdot\sum_{j=s+1}^n z_j \leq u\cdot(1-t) = u\cdot\sum_{j=s+1}^n f_j\cdot z_j.$$

We claim that $\sum_{j=s+1}^n z_j = 0$, i.e., $z_j = 0$ for $s+1 \leq j \leq n$.

For suppose not. Taking into account that $f_j \leq f_i$ $(i \leq s)$ for all j between $s+1$ and n we then obtain

$$u \geq (f_i \cdot \sum_{j=s+1}^{n} z_j) \cdot (\sum_{j=s+1}^{n} f_j \cdot z_j)^{-1} \geq 1,$$

which contradicts Lemma 5.8 and claim follows.

We have thus shown that any $z \in \mathcal{G}_f$ must be of the form $z = (z_1, \ldots, z_s, 0, \ldots, 0)$. Write $z^s = (z_1, \ldots, z_s)$. Evidently, $\|z^s\| = \|z\|$ and $z^s \in (f^s)^{-1}(1)$. In view of Proposition 4.1 and the assumption $q(f) = q(f^s)$ we get

$$q(f) = \|P_{f,z}\| = 1 + \max\{(\|z\| - 2 \cdot z_i) \cdot f_i : 1 \leq i \leq n\} \geq$$

$$1 + \max\{(\|z\| - 2 \cdot z_i) \cdot f_i : 1 \leq i \leq s\} = \|P_{f,z^s}\| \geq q(f^s) = q(f).$$

Consequently, $\|P_{f,z^s}\| = q(f^s)$ and $z \in \mathcal{G}_{f^s}$. ***

Lemma II.5.11. Let $B = 1_1^n$, $n \geq 4$, $f \in S_{B^*}$, $f > 0$. Let $m = m(f) = n$. Let further $z \in \mathcal{G}_f \cap \text{ext } M$, $z \neq z^o$, where

$$z^o = (u/2) \cdot (\beta_n - f_1^{-1}, \ldots, \beta_n - f_n^{-1}).$$

Then

$$\|z\| \leq u \cdot f_n^{-1}. \tag{2.5.6}$$

Proof. It is not hard to check that z^o is the only vertex of M, the set of feasible solutions of the Bf - problem, with all coordinates positive. By our assumption $z \in \mathcal{G}_f$, $z \neq z^o$. Then there exists at least one index $j \leq m = n$ such that $z_j = 0$. By Corollary 4.4 we have

$$z_1 + \ldots + z_{j-1} + z_{j+1} + \ldots + z_n \leq u/f_j. \tag{2.5.7}$$

Hence, $\|z\| \leq u/f_j \leq u/f_n$. ***

Lemma II.5.12. Let $B = 1_1^n$, $n \geq 4$, $f \in S_{B^*}$, $f > 0$. Suppose that $m = m(f) <$ $< k(f) = n$. and let $B_m = 1_1^m$, $f^m = (f_1, \ldots, f_m)$. Then $q(f) = q(f^m)$.

Proof. Since $k(f) \neq m$, we have $\beta_n = f_n^{-1}$. For each $j \in \Gamma(n-m)$, $b_{n-j} = b_{n-j+1} - f_{n-j+1} = (n-j-1) \cdot \beta_{n-j+1} - f_{n-j+1}^{-1} = (n-j-2) \cdot f_{n-j+1}^{-1}$.
By the definition of m we obtain $\beta_{n-j} = \beta_n$, and hence

$$\beta_m = \beta_{m+1} = \ldots = \beta_n. \tag{2.5.8}$$

Further, we have

$$a_n = a_m + \sum_{j=m+1}^{n} f_j = a_m + (n-m)/\beta_m, \text{ and so}$$

$$a_n \cdot \beta_n - n = a_m \cdot \beta_n + (n-m) - n = a_m \cdot \beta_m - m. \tag{2.5.9}$$

Since $k(f) = n$, $a_{n-1} \geq n-3$, whence in view of $a_{n-1} = a_m + (n-m-1) \cdot f_{m+1}$ and $f_{m+1} \leq 1$ we get $a_m \geq m-2$.

It is easy to see that $k(f^m) = m$. (See Lemma 6 of [21]). Thus by Proposition 4.7, $q(f^m) = 1 + 2\cdot(a_m\cdot\beta_m - m)^{-1}$. As observed, $\beta_n = f_n^{-1}$. In view of (2.4.24) and (2.5.9) we finally obtain

$$q(f) = 1 + 2\cdot(a_n\cdot\beta_n - n)^{-1} = 1 + (a_m\cdot\beta_m - m)^{-1} = q(f^m). ***$$

Lemma II.5.13. Let $B = l_1^n$, $n\geq 4$, and let $f \in S_{B^*}$ satisfy (2.5.1). Let $f_n > 0$, $m = m(f) \neq n, B_m = l_1^m$, $f^m = (f_1,\ldots,f_m)$. Then

$$\mathcal{G}_f = \{z \in B^+ : (z_1,\ldots,z_m) \in \mathcal{G}_{f^m},\ z_i = 0 \text{ for } m+1\leq i\leq n\}. \tag{2.5.10}$$

Proof. Let $z^m = (z_1,\ldots,z_m) \in \mathcal{G}_{f^m}$. Write $z = (z_1,\ldots,z_m,0,\ldots,0) \in B$. Then $\|z\| = \|z^m\|$ and $z \in f^{-1}(1)$. On account of Lemma 5.10, we shall be done when we show that $z \in \mathcal{G}_f$. Let

$$z^o = (u/2)\cdot(\beta_m - f_1^{-1},\ldots,\beta_m - f_m^{-1}) \text{ where } 1 + u = q(f).$$

By Lemma 5 12, $q(f) = q(f^m)$.

If $a_m > m-2$, then by Proposition 5.7 card $\mathcal{G}_f = 1$ and by Lemma 5.10 card $\mathcal{G}_f$ = card $\mathcal{G}_{f^m} = 1$, so that (2.5.10) must be true.

If $a_m < m-2$, then by Proposition 5.7 $z^o \notin \mathcal{G}_{f^m}$. Lemma 5.11 with Proposition 4.1 imply

$$\|P_{f,z}\| = 1 + \max\{(\|z\|-2\cdot z_i)\cdot f_i : 1\leq i\leq n\} = 1 + \max\{u, \|z\|\cdot f_{m+1}\} \leq$$

$$\leq 1 + \max\{u, \|z\|\cdot f_m\} = 1 + u = q(f^m) = q(f)$$

and hence $z \in \mathcal{G}_f$.

If $a_m = m-2$, then $\beta_m = \beta_{m+1}$, according to (2.5.9). This yields

$$b_m/(m-2) = (b_m + f_{m+1}^{-1})/(m-1), \text{ whence}$$

$$f_{m+1}\cdot b_m = m-2 = a_m. \tag{2.5.11}$$

Now observe that $\|P_{f,z}\| = 1 + \max\{u, \|z\|\cdot f_{m+1}\}$. If $z^m \neq z^o$, then by Lemma 5.11

$$\|z\| = \|z^m\| \leq u\cdot f_m^{-1} \leq u\cdot f_{m+1}^{-1} \text{ and}$$

$$\|P_{f,z}\| = q(f) = q(f^m).$$

If $z^m = z^o$, then

$$\|z^o\| = (u/2)\cdot\sum_{i=1}^{n}(\beta_m - f_i^{-1}) = (u/2)\cdot(m\cdot\beta_m - b_m) =$$

$$= (u/2)\cdot(m\cdot\beta_m - (m-2)\cdot\beta_m) = u\cdot\beta_m.$$

This together with (2.5.11) gives

$$\|z\|\cdot f_{m+1} = (u\cdot b_m\cdot f_{m+1})/(m-2) = u.$$

Thus again

$$\|P_{f,z}\| = 1+ u = q(f) = q(f^m),$$

showing that $z \in \mathcal{G}_f$, in each case. ***

Proof of the Proposition 5.7. It suffices to consider the most general case, that of $m = m(f) < n$.

Let $\nu = \max\{i; i \in J_f\}$. Then $k(f) \le \nu < n$. and

$f = (f_1, \dots, f_m, \dots, f_{k(f)}, \dots, f_\nu, 0, \dots, 0)$. Write

$f^3 = (f_1, \dots, f_\nu)$, $f^2 = (f_1, \dots, f_{k(f)})$, $f^1 = (f_1, \dots, f_m)$. We have $k(f^1) =$ $= k(f^2) = k(f^3)$ and by Theorem 4.9 $q(f^1) = q(f^2) = q(f^3)$. Finally, applying Corollary 4.5 and Lemmas 5.11, 5.12 we obtain $q(f) = q(f^1)$, and hence

$\mathcal{G}_f = \{z \in B^+ : (z_1, \dots, z_m) \in \mathcal{G}_{f^1},\ z_j = 0 \text{ for } m < j \le n\}$; in particular

card $\mathcal{G}_f$ card $\mathcal{G}_{f^1}$. ***

§ 6. On the uniqueness of minimal projections in c_o and l_1

In this section we deal with spaces c_o and l_1. As it has been shown in [21] (Corollary 1), a minimal projection of c_o onto a subspace $D = f^{-1}(0)$, $f \in S_{(c_o)^*}$, exists provided that either $\|f\|_\infty \ge 1/2$ or $\|f_\infty\| < 1/2$ and the set $J_f = \{i \in \mathbb{N} : f_i \ne 0\}$ is finite.

As regards the space l_1, we confine ourselves to the situation where $D = f^{-1}(0)$, $f \in (l_1)^*$, $f = (f_1, f_2, \dots)$ and

$$1 = f_1 \ge f_2 \ge \dots \ge 0. \tag{2.6.1}$$

In this case (see [21]) a minimal projection onto D does exist and has norm ≥ 1, equality holding only if $f_3 = 0$.

The symbols $P_{f,z}$, $q(f)$, $\mathcal{G}_f$ introduced in § 1 are used also in this section. We shall also need the following

Proposition II.6.1. ([21], Th. 2). Let $B = c_o$ and let $f \in (c_o)^* \cong l_1$, $\|f\|_1 = 1$, $D = f^{-1}(0)$. The norm of a minimal projection onto $f^{-1}(0)$ is strictly greater that one iff

(i) $\|f\|_\infty < 1/2$;

(ii) the set $J_f = \{i \in \mathbb{N} : f_i \ne 0\}$ is finite.

In that case,

$$q(f) = 1 + (\sum_{i=1}^{\infty} |f_i|/(1-2\cdot|f_i|))^{-1}. \qquad (2.6.2)$$

If, moreover, $z \in f^{-1}(1)$, then

$$\|P_{f,z}\| = \sup(|1-f_i\cdot z_i| + |z_i|\cdot(1-|f_i|): i \in \mathbb{N}) \qquad (2.6.3)$$

A minimal projection P onto $f^{-1}(0)$ ($\|P\|=1, \|f\|_1=1$) is unique iff $|f_i| \geq 1/2$ for exactly one index i.

Theorem II.6.2. Let $B = c_o$, $f \in S_{B^*}$, $\|f\| = 1, q(f) > 1$. Then card $\mathcal{G}_f > 1$.

Proof. The assumptions force the finitness of J_f, in view of Proposition 6.1. Let $i_o > \max J_f$. Consider the space $B_o = l_{\infty}^{i_o}$. Let $f^o = (f_1,\ldots,f_{i_o})$.

In virtue of Theorem 3.6 the subspace $D_o = (f^o)^{-1}(0)$ admits at least two distinct minimal projections P_1 and P_2 from B_o.

Suppose that $P_1 = P_{f^o,z^{1o}}$, $P_2 = P_{f^o,z^{2o}}$, where $z^{1o}, z^{2o} \in (f^o)^{-1}(1)$, $z^{1o} \neq z^{2o}$. Let $z^1 = (z_1^{1o}, z_2^{1o},\ldots,z_{i_o}^{1o},0,\ldots) \in c_o$, $z^2 = (z_1^{2o}, z_2^{o2},\ldots,z_{i_o}^{2o},0,\ldots) \in c_o$. Obviously, $z^1 \neq z^2$, $z^1, z^2 \in f^{-1}(1)$, and hence, by Proposition 3.1 and Proposition 6.1,

$$\|P_{f,z^1}\| = \|P_{f,z^2}\| = \|P_{f^o,z^{1o}}\| = \|P_{f^o,z^{2o}}\| = q(f) = q(f^o). ***$$

Proposition II.6.3. ([21],Th. 7). Let $B = l_1$, $f \in S_{B^*}$, $1=f_1\geq f_2\geq\ldots\geq 0$, $f_3 > 0$. If $f = f^o = (1,1,\ldots)$ then $q(f) = 2$. If $f \neq f^o$ then there exist $c_j = \min(f_j\cdot b_{j-1}, a_{j-1})$ and $k = k(f) = \max(j: c_j - j \geq -3)$, where b_j, a_j are defined by (2.4.16), and we have $q(f) = 1 + u$, where u is expressed by formula (2.4.24).

Proposition II.6.4. ([21],Lemma 4). Let $B = l_1$, $f \in S_{B^*}$, $f \geq 0$. Then

$$\mathcal{G}_f = \{z \in B: z \in f^{-1}(1), z \geq 0, \|P_{f,z}\| = q(f)\} \qquad (2.6.4)$$

Theorem II.6.5. Let $B = l_1$, $f \in S_{B^*}$ and suppose $1=f_1\geq f_2\geq\ldots\geq 0$, $f_3>0$.

(a) If $f = f^o= (1,1,\ldots)$ then card $\mathcal{G}_f > 1$.

(a) If $f \neq f^o$, consider $B_m = l_1^m$ with m given by (2.5.2) and let $f^m = (f_1,\ldots,f_m)$. A minimal projection from B onto $f^{-1}(0)$ is unique iff a minimal projection from B_m onto $(f^m)^{-1}(0)$ is unique.

Proof. (a) Let $f = f^o$, $z^1= (1/2,1/4,1/4,0,\ldots) \in B$.

$z^2 = (1/3,1/3,1/3,0,\dots) \in B$. Clearly, $z^1, z^2 \in f^{-1}(1)$ and $P_{f^o,z^1} \neq P_{f^o,z^2}$. In view of (2.6.3) and Proposition 6.3 we have $\|P_{f^o,z^1}\| = \|P_{f^o,z^2}\| = q(f) = 2$. Hence card $\mathcal{G}_{f^o} > 1$.

(b) Let $f \neq f^o$. By Proposition 6.3 and Lemma 5.8, $q(f) = 1 + u < 2$; it suffices just to consider $B_k = l_1^k$, $f^k = (f_1,\dots,f_k)$, where $k = k(f)$ is given by (2.4.23), and notice that $q(f^k) = q(f)$.

Let $m = m(f)$ be given by (2.5.2), let $B_m = l_1^m$. and $f^m = (f_1,\dots,f_m)$. Then $q(f^m) = q(f)$ by Proposition 6.3 and 5.7.

Write $B^+ = \{z \in B: z_i \geq 0\}$. We will show that

$$\mathcal{G}_f \subset \{z \in B^+ : (z_1,\dots,z_m) \in \mathcal{G}_{f^m},\ z_i = 0 \text{ for } i>m\} \tag{2.6.5}$$

Let $z \in \mathcal{G}_f$ and $z^m = (z_1,\dots,z_m)$. By (2.6.4), $z \in B^+$.

Assume that there exists an index $i > m$ such that $z_i > 0$. Then $\|z^m\| < \|z\|$.

First examine the case $f_{m+1} = 0$ (and hence by assumption, $f_{m+1} = f_{m+2} = \dots = 0$). then $z^m \in (f^m)^{-1}(1)$. According to Proposition 6.4 and Theorem 4.9,

$$q(f^m) \leq \|P_{f^m,z^m}\| = 1 + \max\{f_i\cdot(\|z^m\|-2\cdot z_i): 1\leq i\leq m\} \leq$$

$$\leq 1 + \max\{f_i\cdot(\|z\|-2\cdot z_i): 1\leq i\leq m\} = \|P_{f,z}\| = q(f)$$

Consequently $\|z\| \neq \|z^m\|$, contradicting the assumption that $z_i > 0$ for $i>m$, and $z^m \in \mathcal{G}_{f^m}$.

Now consider the case $f_{m+1} \neq 0$. We claim that $z_1 > 0$. Indeed; $\|z\| = \|z\|\cdot\|f\| \geq f(z) = 1$, and by Proposition 6.3

$\|z\| - 2\cdot z_1 = (\|z\|-2\cdot z_1)\cdot f_1 \leq \|P_{f,z}\| - 1 = q(f) - 1 = u < 1$, whence the claim. Therefore $t = \sum_{i=1}^{m} f_i\cdot z_i > 0$.

Let $y^m = t^{-1}\cdot z$. Clearly $y^m \in (f^m)^{-1}(1)$. By Proposition 6.3 there is an index i_o, $1\leq i_o<m$, for which

$\|P_{f^m,y^m}\| = 1 + f_{i_o}\cdot(\|y^m\|-2\cdot y_{i_o})$. Hence

$$t\cdot u \leq t\cdot(\|P_{f^m,y^m}\| - 1) = t\cdot f_{i_o}\cdot(\|y^m\|-2\cdot y_{i_o}) = f_{i_o}\cdot(\|t\cdot y^m\|-2\cdot z_{i_o}) =$$

$$f_{i_o}\cdot(\|z\| - \sum_{j=m+1}^{\infty} z_j - 2\cdot z_{i_o}) = f_{i_o}\cdot(\|z\|-2\cdot z_{i_o}) - f_{i_o}\cdot\sum_{j=m+1}^{\infty} z_j \leq$$

$$\leq u - f_{i_o}\cdot\sum_{j=m+1}^{\infty} z_j \leq u\cdot(1-t) = u\cdot\sum_{j=m+1}^{\infty} f_j\cdot z_j .$$

Since $f_{i_o} \geq f_j$ for all $j \geq m+1$, we get $u \geq 1$, a contradiction. Our assumption ($z_i > 0$ for some $i > m$) must be false.

Thus $\|P_{f^m,z^m}\| = \|P_{f,z}\| = q(f) = q(f^m)$, showing that $z^m \in \mathcal{G}_{f^m}$.

This establishes inclusion (2.6.5). It remains to prove the opposite inclusion, to conclude that

$$\mathcal{G}_f = \{z \in B_+ : (z_1,\dots,z_m) \in \mathcal{G}_{f^m},\ z_i = 0 \text{ for } i > m\} \qquad (2.6.6)$$

Choose $z^m \in \mathcal{G}_{f^m}$, $z^m = (z_1^m,\dots,z_m^m)$, and define $z = (z_1^m,\dots,z_m^m,0,\dots) \in B$. According to Lemma 4.2, $z^m \in (B_m)_+$, whence $z \in f^{-1}(1) \cap B_+$. By Theorem 4.9 and condition (2.6.1), $\|P_{f,z}\| = 1 + \max\{u, f_{m+1}\cdot\|z\|\}$.

If $f_{m+1} = 0$ then

$\|P_{f,z}\| = \|P_{f^m,z^m}\| = q(f^m) = q(f)$,

in view of Theorem 4.9 and Proposition 5.7, and so $z \in \mathcal{G}_f$.

If $f_{m+1} > 0$, we define

$f^{m+1} = (f_1,\dots,f_{m+1})$, $B_{m+1} = l_1^{m+1}$, $z^{m+1} = (z_1^m,\dots,z_m^m,0)$

and apply Proposition 5.7 to obtain

$\mathcal{G}_{f^{m+1}} = \{x \in (B_{m+1})_+ : (x_1,\dots x_m) \in \mathcal{G}_{f^m},\ x_{m+1} = 0\}$ and

$\|P_{f^{m+1},z^{m+1}}\| = \|P_{f^m,z^m}\| = q(f^m) = q(f) = 1 + u$.

Since $\|P_{f^{m+1},z^{m+1}}\| = 1 + \max\{u, f_{m+1}\cdot\|z^{m+1}\|\}$, comparing of the right side yields $f_{m+1}\cdot\|z^{m+1}\| \leq u$.

Finally, observe that $\|z^{m+1}\| = \|z^m\| = \|z\|$. Consequently,

$\|P_{f,z}\| = 1 + \max\{u, f_{m+1}\cdot\|z\|\} = 1 + u = q(f)$ and so $z \in \mathcal{G}_f$. ***

§ 7. Of infimum of norms of projections on subspaces codimension one.

Let B be a Banach space. Set

$$\Delta_1(B) = \sup\{\rho(B,D) : D \text{ is a closed hyperplane in } D\} \qquad (2.7.1)$$

The main problem of this section is to estimate the constant $\Delta_1(B)$ in case $B = L^p([0,1],\mu)$ when μ is the Lebesgue measure (We will write for brevity L^p) and in case $B = l^p$, where $1 < p < +\infty$.

Recall that if $P \in \mathcal{P}(B,D)$ where $D \subset B$ is a hyperplane (we may assume $D = \ker f$ for some $f \in S_{B^*}$) then there exists $y_p \in B$, $f(y_p) = 1$ such that P is of the form

$$Px = x - f(x)\cdot y_p \text{ for every } x \in B. \tag{2.7.2}$$

Conversely, if $y \in B$ satisfies $f(y) = 1$ then the operator P_y defined by

$$P_y x = x - f(x)\cdot y \text{ for } x \in B \tag{2.7.3}$$

is a projection from B onto D (see [21]).

We note that a trivial estimation gives

$$\|P\| \le 1 + \|y_p\|. \tag{2.7.4}$$

Since for every $\varepsilon > 0$ we can find $y \in B$ satisfying $f(y) = 1$ and $\|y\| < 1+\varepsilon$, we obtain $\rho(B,D) \le 2$ for every hyperplane $D \subset B$ and consequently

$$\Delta_1(B) \le 2 \tag{2.7.5}$$

The main result of this section is to prove the following estimation:

$$\Delta_1(l^p) \le \Delta_1(L^p) \le 2^{|2/p-1|} \tag{2.7.6}$$

for $1 < p < +\infty$.

The proof is based on the following

Proposition II.7.1. Let D_1, D_2 be two subspaces of codimension two in B. Suppose that there is an isometry T of B onto itself such that

$$T(D_1) = T(D_2) \tag{2.7.7}$$

Then

$$\rho(B,D_1) = \rho(B,D_2). \tag{2.7.8}$$

Proof. Let P_1 be an arbitrary projection mapping B onto D_1. Then

$$P_2 = T\circ P_1\circ T^{-1} \tag{2.7.9}$$

is a projection mapping B onto D_2. Since the operator T is an isometry, $\|T^{-1}\| = 1$. Therefore, by (2.7.9), $\|P_2\| \le \|T\|\cdot\|P_1\|\cdot\|T^{-1}\|$ and it implies $\rho(B,D_2) \le \rho(B,D_1)$.

Changing the role of D_1 and D_2 we obtain equality (2.7.8).***

Corollary II.7.2. Let $f_1, f_2 \in S_{B^*}$. Suppose that there is an isometry T of B onto itself such that the conjugate isometry T^* maps f_1 onto f_2. Then $\rho(B,\ker f_1) = \rho(B,\ker f_2)$.

Proof. It is trivial that if $T^*f_1 = f_2$ then $T(\ker f_2) = \ker f_1$.***

Corollary II.7.3. (see [151]) Suppose that B is a reflexive space. Suppose that for two arbitrary continuous linear functionals $f_1,f_2 \in S_{B^*}$ there is an isometry T^* mapping B^* onto itself such that $T^*f_1 = f_2$. Then for all subspaces D of codimension one the numbers $\rho(B,D)$ are equal.

Proof. In the representation of D we may assume D = kerf for some $f \in S_B*$. Then we apply Corollary 7.2.***

An example of a measure μ such that the space $L^p(\Omega,\Sigma,\mu)$, $1 < p < +\infty$ satisfies the assumption of Corollary 7.3 is given in [167] ,Prop.IX.6.7. Of course, for $p \neq 2$ these spaces are not Hilbert spaces and it was observed in [151] that $\rho(B,D)$ are equal for all D. The space B, in which for arbitrary x_1, x_2 of norm one there is an isometry T such that $Tx_1=x_2$ is called isotropic. Then Corollary 7.3 may be reformulated as

Corollary II.7.4. Let B be a reflexive space. If B^* is isotropic then $\rho(B,D)$ are equal for all subspaces D of codimension one.

We shall say that a Banach space B is almost isotropic if for arbitrary $x_1,x_2 \in S_B$ and arbitrary $\varepsilon>0$ there is an isometry T_ε mapping B into itself that

$$\|T_\varepsilon x_1 - x_2\| < \varepsilon \tag{2.7.10}$$

It was shown ([167],Th.IX.6.4) that the space L^p is almost isotropic.

Theorem II.7.5. Let B be a reflexive Banach space. If the space B^* is almost isotropic, then $\rho(B,D)$ are equal for all subspaces D of codimension one.

The proof is based on the following simple

Lemma II.7.6. The function $\rho(B,\ker(f))$ is a locally Lipschitz function of f on S_B*.

Proof. Let $f \in S_B*$. Let ε be an arbitrary positive number less that 1.

Select $P \in \mathcal{P}(B,\ker(f))$ with

$$\|P\| \le \rho(B,\ker(f)) + \varepsilon \tag{2.7.11}$$

Following (2.7.2) P is of the form

$$Px = x - f(x)\cdot y_P \tag{2.7.12}$$

where $f(y_P) = 1$. Since $f(\cdot)\cdot y_P = I - P$, by (2.7.5),

$$\|y_P\| \le \|P\| + \|I\| \le \rho(B,\ker(f)) + 1 + \varepsilon \le 3 + \varepsilon < 4. \tag{2.7.13}$$

Select $f_o \in S_B*$ with $\|f - f_o\| \le 1/8$. Then , of course, by (2.7.13),

$$f_o(y_P) = f(y_P) + (f_o(y_P) - f(y_P)) \ge 1- \|f - f_o\|\cdot\|y_P\| \ge 1/2. \tag{2.7.14}$$

Put $\beta = 1/f_o(y_P)$ and $g_o = \beta\cdot f_o$. Then, of course, $g_o(y_P) = 1$ and the operator $P_o = I - g_o(\cdot)\cdot y_P$ belongs to the set $\mathcal{P}(B,\ker(f_o))$.

Now we shall estimate

$$\|P - P_o\| \le \|y_P\|\cdot\|f-g_o\| = \|y_P\|\cdot(\|f-f_o\|+|\beta-1|\cdot\|f_o\|) = \|y_P\|\cdot(\|f-f_o\|+|\beta-1|).$$

Now the essential role plays the estimation of $|\beta-1|$. Recall that $f(y_P)=1$

and $f_o(y_p) = 1/\beta$. Hence

$$|(\beta-1)/\beta| = |f(y_p) - f_o(y_p)| \le \|y_p\|\cdot\|f - f_o\|.$$

It is easy to check that $|\beta| \le 2$. Thus we obtain that

$$|\beta - 1| \le 2\cdot\|y_p\|\cdot\|f - f_o\| \text{ and consequently}$$

$$\|P - P_o\| \le K\cdot\|f - f_o\|, \text{ where } K = \|y_p\|\cdot(1 + 2\cdot\|y_p\|) = 36.$$

Then, following (2.7.11), $\rho(B,\ker(f_o)) \le \rho(B,\ker(f)) + K\cdot\|f - f_o\|$.

Changing the role f and f_o we can obtain a converse inequality. This implies that $\rho(B,\ker(f))$ is locally Lipschitzian.***

Proof of Theorem II.7.5. Since B^* is almost isotropic then $\rho(B,\ker(f))$ is constant on a dense subset of S_{B^*}. By Lemma 2.7.6 it is continuous. Thus it is constant on the whole S_{B^*}.***

Corollary II.7.7. The constants $\rho(L^p,D)$ are equal for all subspaces D of codimension one.

Now we are able to prove the following

Theorem II.7.8. There holds the following estimation

$$\Delta_1(L^p) \le 2^{|2/p-1|} \text{ for } 1 < p < +\infty.$$

Proof. Since, by Corollary 7.7, $\rho(L^p,D)$ are equal for all subspaces D of codimension one, we can select the most convenient subspace for our considerations. It will be a subspace $D = \ker(f)$, where

$$f(x) = \int_0^1 x(\tau)\, d\tau. \tag{2.7.15}$$

Let P be an operator on D defined by

$$Px = x - f(x)\cdot 1 \tag{2.7.16}$$

(The symbol 1 denotes the function $h(t) = 1$ for every $t \in [0,1]$.)
It is easy to check that $P \in \mathcal{P}(B,D)$. Hence $\Delta_1(L^p) \le \|P\|$.

Observe that P acts on all spaces L^p, $1 \le p \le +\infty$. Denote by $q(p)$ the norm of P in the space L^p. It is easy to calculate that $q(1) = q(+\infty) = 2$, $q(2) = 1$. Recall that, following M. Riesz interpolation theorem (see [165]), $\ln(q(1/s))$ is a convex function of s, $0<s\le1$. Hence we trivially obtain $q(p) \le 2^{|2/p-1|}$ which finishes the proof.***

Theorem II.7.9. The following equality is true

$$\rho(L^p[0,+\infty),D) = \Delta_1(L^p) \text{ ,} 1 < p < +\infty,$$

for all subspaces $D \subset L^p[0,+\infty)$ of codimension one.

Proof. Let F be a functional generated by the function with finite support. By the previous considerations $\rho(L^p[0,+\infty),\ker(F)) = \Delta_1(L^p)$.

The functionals generated by the functions of finite support are dense. Thus, by Lemma 7.6, $\rho(L^p[0,+\infty),D) = \Delta_1(L^p)$ for all subspaces D of codimension one.***

Theorem II.7.10. There holds the following estimation

$$\Delta_1(l^p) \le \Delta_1(L^p) \text{ for } 1 < p , +\infty.$$

Proof. The space l^p can be embedded in $L^p[o,+\infty)$ in the following way

$$x(t) = I(\langle x_n \rangle) = x_n \text{ for } n-1 \le t < n, \text{ where } \langle x_n \rangle \in l^p.$$

Let F be an arbitrary linear continuous functional defined on l^p. We can extend F with preserving of the norm on the whole space $L^p[0,+\infty)$ as a functional generated by a function $f(t)$ which is constant on the intervals $[n,n+1)$, $n=0,1,\dots$.

Let $P_o \in \Delta(L^p[0,+\infty),\ker(F))$. The difficulty is that P_o may map $I(l^p)$ not into itself. Thus we construct another minimal projection having the reqired property. For this purpose we construct a family of isometries T_s, $0 \le s < 1$, defined in the following way

$$(T_s x)t = \begin{cases} x(t+s) & \text{if } E(t+s) = E(t) \\ x(t+s-1) & \text{if } E(t+s) > E(t) \end{cases}$$

where $E(a)$ denotes the integer part of a. Let

$$Px = \int_0^1 (T_s \circ P_o \circ T_s^{-1})x \, ds.$$

We shall show that P is a projection. Recalling that f is constant on intervals $[n,n+1)$, $n=0,1,\dots$, we obtain that

$$f(T_s x) = f(x) = f(T_s^{-1} x) \text{ for } 0 \le s < 1.$$

Then the operator P is of the form $Px = x - f(x)\cdot y$, where

$$y(t) = \int_0^1 (T_s y_o)t \, dt$$

is constant on intervals $[n,n+1)$ and such that

$$y(n) = \int_0^1 y_o(s) \, ds.$$

(The function y_o is so chosen that $P_o = I - F(\cdot)\cdot y_o$.) It implies that $f(y) = 1$ and P is a projection. Moreover, if $x \in l^p$, the form of P implies that $Px \in l^p$. Observe that $\|P\| \le \int_0^1 \|P_o\| \, ds = \|P_o\|$, since $\|T_s\| = \|T_s^{-1}\| = 1$. Hence $\Delta_1(l^p) \le \Delta_1(L^p[0,+\infty)) \le \Delta_1(L^p)$ which gives the result.***

§ 8. On projections of subspaces of finite codimension

Let B be a Banach space and let $D \subset B$ be its subspace. Let us introduce the following notations

$$\bar{\Delta}_k(B) = \sup\{\rho(B,D): D \subset B,\ \operatorname{codim} D = k\} \tag{2.8.1}$$

and $$\underline{\Delta}_k(B) = \inf\{\rho(B,D): D \subset B,\ \operatorname{codim} d = k\} \tag{2.8.2}$$

In the previous section it was shown that for spaces L^p we have

$$\underline{\Delta}_1(L^p) = \bar{\Delta}_1(L^p) \le 2^{|2/p-1|} \text{ if } 1 < p < +\infty.$$

Thus a natural question arises about the extention of the above equality for $k > 1$. In this section it will be shown that

$$\underline{\Delta}_k(L^p) \le \Delta_1(L^p) \tag{2.8.3}$$

and $$\bar{\Delta}_2(L^p) > \Delta_1(L^p) \tag{2.8.4}$$

for p either close to 1 or sufficiently large.
We start with the following

Theorem II.8.1. For $k \ge 2$ the inequality (2.8.3) holds true.

Proof. Let an integer $k > 1$ be fixed. Let us define

$$D = \{x \in L^p: \int_{(i-1)/k}^{i/k} x(t)\,dt = 0,\ i=1,\dots,k\}. \tag{2.8.5}$$

and for $i=1,\dots,k$ put

$$B_i = \{x \in L^p: \operatorname{supp} x \subset [(i-1)/k, i/k]\}. \tag{2.8.6}$$

It is easy to check that L^p is a direct sum of $B_1,\dots,B_k$ and for $x_i \in B_i$ $(i=1,\dots,k)$ we have $\|x_1+\dots+x_k\| = (\|x_1\|^p+\dots+\|x_k\|^p)^{1/p}$.
Observe that $D \cap B_i$ is a subspace of codimension 1 in B_i. Thus for each $\varepsilon>0$ there is a projection P_i mapping B_i onto $D \cap B_i$ with the norm $\|P_i\| \le \Delta_1(L^p) + \varepsilon$. Let $Px = P_1x_1+\dots+P_kx_k$ for $x = x_1+\dots+x_k$, $x_i \in B_i$.
Then P is a projection from B onto D and

$$\|Px\| = (\|P_1x_1\|^p+\dots+\|P_kx_k\|^p)^{1/p} \le (\Delta_1(L^p)+\varepsilon)\cdot(\|x_1\|^p+\dots+\|x_k\|^p)^{1/p} \le$$
$$\le (\Delta_1(L^p)+\varepsilon)\cdot\|x\|.$$

The arbitrarness of ε implies that $\underline{\Delta}_k(L^p) \le \Delta_1(L^p)$ which gives the result.***

The proof of inequality (2.8.4) is more complicated and needs some preliminary results.
At first let us define

$$D = \{x \in L^p: \int_0^1 \sin(2\pi t x(t))\, dt = 0 \text{ and } \int_0^1 \cos(2\pi t x(t))\, dt = 0. \quad (2.8.7)$$

Clearly D is a subspace of L^p, $1 < p < +\infty$ and codim D = 2.

Proposition II.8.2. The linear operator

$$Px = x - 2\cdot(\int_0^1 \sin(2\pi\tau)\cdot x(\tau)\, d\tau \cdot \sin(2\pi t) + \quad (2.8.8)$$

$$+ \int_0^1 \cos(2\pi\tau)\cdot x(\tau)\, d\tau\cdot\cos(2\pi t))$$

mapping L^p onto D is a projection with minimal norm.

Proof. Let T_s, $0 \le s < 1$, be a family of isometries mapping L^p onto itself and defined as follows

$$(T_s x)t = \begin{cases} x(t+s) & \text{if } t+s \le 1 \\ x(t+s-1) & \text{if } t+s > 1 \end{cases} \quad (2.8.9)$$

Observe that the space D is invariant under T_s. Indeed,

$$\int_0^1 \sin(2\pi t)\cdot(T_s x)t\, dt = \int_0^1 \sin(2\pi(t-s))\cdot x(t)\, dt =$$

$$\int_0^1 \sin(2\pi t)\cdot\cos(2\pi s)\cdot x(t)\, dt - \int_0^1 \cos(2\pi t)\cdot\sin(2\pi s)\cdot x(t)\, dt = 0$$

by the definition of D. In a similar way we can prove that

$$\int_0^1 \cos(2\pi t)\cdot(T_s x)t\, dt = 0,$$

and consequently $T_s D = D$ for each $s \in [0,1)$.

Now we shall show that that the operator P defined by (2.8.7) belongs to $\mathcal{P}(L^p,D)$. To begin with, we observe that linear combinations of elements of the set $E = \{1, \sin(2\cdot 2\pi t), \cos(2\cdot 2\pi t), \ldots, \sin(k\cdot 2\pi t), \cos(k\cdot 2\pi t), \ldots\}$ are dense in the space D. Fix an arbitrary $P_o \in \mathcal{P}(B,D)$. Let us set

$$P_1 = \int_0^1 T_s \circ P_o \circ T_s^{-1}\, ds \quad (2.8.10)$$

We show that $P_1 \in \mathcal{P}(B,D)$. To do this, note that

$$P_1 \sin(2k\pi t) = \int_0^1 (T_s \circ P_o \circ T_s^{-1}) \sin(2k\pi t)\, ds = \int_0^1 (T_s \circ P_o) \sin(2k\pi(t-s))\, ds =$$

$$= \int_0^1 (T_s \circ P_o)(\sin(2k\pi t)\cdot\cos(2k\pi s) - \cos(2k\pi t)\cdot\sin(2k\pi s))\, ds =$$

$$= \int_0^1 T_s(\sin(2k\pi t)\cdot\cos(2k\pi s) - \cos(2k\pi t)\cdot\sin(2k\pi s))\, ds =$$

$$= \int_0^1 (T_s \circ T_s^{-1}) \sin(2k\pi t)\, ds = \sin(2k\pi t).$$

By a similar consideration $P_1 \cos(2k\pi t) = \cos(2k\pi t)$. Since the linear combinations of the set E are dense in D, we get $P_1 x = x$ for every $x \in D$. Now we calculate $P_1 x$ more precisely. Recall that P_o can be represented in in the form (see e.g. [21]),

$$P_o x = x - \int_0^1 \sin(2\pi\tau)\cdot x(\tau)\, d\tau \cdot x_s - \int_0^1 \cos(2\pi\tau)\cdot x(\tau)\, d\tau \cdot x_c, \tag{2.8.11}$$

where x_s and x_c are such that

$$\int_0^1 \sin(2\pi t)\cdot x_s(t)\, dt = 1 = \int_0^1 \cos(2\pi t)\cdot x_c(t)\, dt \tag{2.8.12}$$

$$\int_0^1 \sin(2\pi t)\cdot x_c(t)\, dt = 0 = \int_0^1 \sin(2\pi t)\cdot x_s(t)\, dt.$$

Let P_2 be a projection defined as follows:

$$P_2 x = \int_0^1 \sin(2\pi\tau)\cdot x(\tau)\, d\tau \cdot x_s + \int_0^1 \cos(2\pi\tau)\cdot x(\tau)\, d\tau \cdot x_c. \tag{2.8.13}$$

Set

$$P_3 x = \int_0^1 (T_s \circ P_2 \circ T_s^{-1}) x\, ds. \tag{2.8.14}$$

We show that $P = I - P_3$, where P is given by (2.8.7). Compute

$$P_3 x = \int_0^1 \left(\int_0^1 \sin(2\pi(\tau+s))\cdot x(\tau)\, d\tau\right)\cdot (T_s x_s)\, ds +$$

$$+ \int_0^1 \left(\int_0^1 \cos(2\pi(\tau+s))\cdot x(\tau)\, d\tau\right)\cdot (T_s x_c)\, ds =$$

$$= \int_0^1 \left(\int_0^1 (\sin(2\pi\tau)\cdot\cos(2\pi s)+\cos(2\pi\tau)\cdot\sin(2\pi s))\cdot x(\tau)\, d\tau\right)\cdot (T_s x_s)\, ds +$$

$$+ \int_0^1 \left(\int_0^1 (\cos(2\pi\tau)\cdot\cos(2\pi s)-\sin(2\pi\tau)\cdot\sin(2\pi s))\cdot x(\tau)\, d\tau\right)\cdot (T_s x_c)\, ds =$$

$$= \int_0^1 \left(\int_0^1 \sin(2\pi\tau)\cdot x(\tau)\, d\tau\right)\cdot\cos(2\pi s)\cdot (T_s x_s)\, ds +$$

$$+ \int_0^1 \left(\int_0^1 \cos(2\pi\tau)\cdot x(\tau)\, d\tau\right)\cdot\sin(2\pi s)\cdot (T_s x_s)\, ds +$$

$$+ \int_0^1 \left(\int_0^1 \cos(2\pi\tau)\cdot x(\tau)\, d\tau\right)\cdot\cos(2\pi s)\cdot (T_s x_c)\, ds -$$

$$- \int_0^1 \left(\int_0^1 \sin(2\pi\tau)\cdot x(\tau)\, d\tau\right)\cdot\sin(2\pi s)\cdot (T_s x_c)\, ds =$$

$$= \int_0^1 \sin(2\pi\tau)\cdot x(\tau)\, d\tau \cdot \int_0^1 \cos(2\pi(u-\cdot))\cdot x_s(u)\, du +$$

$$+ \int_0^1 \cos(2\pi\tau)\cdot x(\tau)\, d\tau \cdot \int_0^1 \sin(2\pi(u-\cdot))\cdot x_s(u)\, du +$$

$$+ \int_0^1 \cos(2\pi\tau)\cdot x(\tau)\, d\tau \cdot \int_0^1 \cos(2\pi(u-\cdot))\cdot x_c(u)\, du -$$

$$- \int_0^1 \sin(2\pi\tau)\cdot x(\tau)\, d\tau \cdot \int_0^1 \sin(2\pi(u-\cdot))\cdot x_c(u)\, du =$$

$$= \int_0^1 \sin(2\pi\tau)\cdot x(\tau)\, d\tau \cdot \int_0^1 (\sin(2\pi u)\cdot \sin(2\pi\cdot) - \cos(2\pi\cdot)\cdot\cos(2\pi u))\cdot x_s(u)\, du +$$

$$+ \int_0^1 \cos(2\pi\tau)\cdot x(\tau)\, d\tau \cdot \int_0^1 (\sin(2\pi u)\cdot \cos(2\pi\cdot) - \cos(2\pi u)\cdot\sin(2\pi\cdot))\cdot x_s(u)\, du +$$

$$+ \int_0^1 \cos(2\pi\tau)\cdot x(\tau)\, d\tau \cdot \int_0^1 (\cos(2\pi u)\cdot \cos(2\pi\cdot) + \sin(2\pi u)\cdot\sin(2\pi\cdot))\cdot x_c(u)\, du -$$

$$- \int_0^1 \sin(2\pi\tau)\cdot x(\tau)\, d\tau \cdot \int_0^1 (\sin(2\pi u)\cdot \cos(2\pi\cdot) + \cos(2\pi u)\cdot\sin(2\pi\cdot))\cdot x_c(u)\, du =$$

$$= \int_0^1 \sin(2\pi\tau)\cdot x(\tau)\, d\tau \cdot \sin(2\pi\cdot) + \int_0^1 \cos(2\pi\tau)\cdot x(\tau)\, d\tau\cdot\cos(2\pi\cdot).$$

Following (2.8.7), $P = I - P_s$. By (2.8.10),(2.8.13) and (2.8.14),

$= \int_0^1 (T_s \circ P_o \circ T_s^{-1})\, ds$. Consequently $\|P\| \le \|P_o\|$, since $\|T_s\| = \|T_s^{-1}\| = 1$.

By arbitrarness of P_o, $P \in \Delta(B,D)$ which completes the proof.***

Proposition II.8.3. In the space L^1 the operator P given by (2.8.7) has the norm no less than $1 + 4/\pi$.

Proof. First we shall show that the norm of the projection P_s defined by (2.8.14) is equal to $4/\pi$. Let ε be an arbitrary positive number. Let $\delta>0$ be chosen in such a way that $|\sin(2\pi t) - \sin(\pi/4)| < \varepsilon$ and $|\cos(2\pi t) - \cos(\pi/4)| < \varepsilon$ for $|t - 1/4| < \delta$. Define

$$x_\varepsilon(t) = \begin{cases} 1/2\delta & \text{if } |t - 1/4| \le \delta \\ 0 & \text{if } |t - 1/4| > \delta \end{cases} \tag{2.8.15}$$

It is easy to check that $\|x_\varepsilon\| = 1$. Observe that

$\int_0^1 \sin(2\pi\tau)\cdot x_\varepsilon(\tau)\, d\tau = \int_0^1 \cos(2\pi\tau)\cdot x_\varepsilon(\tau)\, d\tau > 2^{-1/2} - \varepsilon$. Hence

$$\|P_s x_\varepsilon\| = 2\cdot a\cdot \int_0^1 |\sin(2\pi t) + \cos(2\pi t)|\, dt =$$

$$= 2\cdot a\cdot (\int_0^{3/8} (\sin(2\pi t) + \cos(2\pi t))\, dt + \int_{7/8}^{0} (\sin(2\pi t) + \cos(2\pi t))\, dt -$$

$$- \int_{3/8}^{7/8} (\sin(2\pi t) + \cos(2\pi t))\, dt) = 4\cdot a\cdot \int_{-1/8}^{3/8} (\sin(2\pi t) + \cos(2\pi t))\, dt =$$

$$= 4\cdot a\cdot 2^{3/2}/2\pi > 4\cdot 2^{1/2}\cdot(2^{-1/2} - \varepsilon)/\pi = 4/\pi - 4\cdot 2^{1/2}\cdot\varepsilon/\pi.$$

(Here $a = \int_0^1 \sin(2\pi t)\cdot x_\varepsilon(t)\, dt = \int_0^1 \cos(2\pi t)\cdot x_\varepsilon(t)\, dt$.) The arbitrarness of ε implies that $\|P_s\| \ge \pi/4$. By Babenko-Pricugov theorem [9]), we get $\|P\| \ge 1 + \pi/4$.***

Theorem II.8.4. The following inequality holds for p sufficiently close to 1: $\|P\|_p > 2$, where P is defined by (2.8.7) (the symbol $\|P\|_p$ denotes the operator norm with respect to L^p space).

Proof. Note that $\|x_\varepsilon\|_p$ and $\|Px_\varepsilon\|$, where x_ε is given by (2.8.15) are continuous functions of p. Since $\|Px_\varepsilon\|_1/\|x_\varepsilon\|_1 \geq 1+4/\pi > 2$, we get the theorem.***

Theorem II.8.5. $\|P\|_q > 2$ for q sufficiently large.

Proof. By the form of P (see (2.8.7)), the operator P^* conjugate to P is of the same form and $\|P^*\| = \|P\|$.***

Finally we obtain

Theorem II.8.6. The following inequality holds for p either sufficiently close to 1 or sufficiently large:

$$\bar{\Delta}_2(L^p) > \Delta_1(L^p) \quad (k\geq 2) \qquad (2.8.16)$$

At the end of this section we consider the following inequality

$$\bar{\Delta}_k(D) \leq \bar{\Delta}_k(B) \text{ for D being a subspace of B.} \qquad (2.8.17)$$

In general, it is not known if (2.8.16) holds true. We shall show that (2.8.16) is valid for B being the Orlicz space of functions defined on the interval $[0,+\infty]$ and D being the corresponding sequence Orlicz space.

Now we introduce some notations. Let N be a convex increasing function mapping $[0,+\infty)$ into itself and such that $N(0) = 0$. We assume that $N(2u) \leq k\cdot N(u)$. By $L_N[0,+\infty)$ we shall denote the space of all measurable functions x such that there exists a number k_x with the property

$$\int_0^{+\infty} N(|x(t)/k_x|)\, dt < +\infty \qquad (2.8.18)$$

Similarly by l^N we shall denote the space of all sequences $x = \{x_n\}$ such that there exists k_x with the property

$$\sum_{n=0}^{\infty} N(|x_n/k_x|) < +\infty \qquad (2.8.19)$$

The spaces $L_N[0,+\infty)$ and l^N are called Orlicz spaces. For the norms and other properties of Orlicz spaces see for example [108],[136]. Of course there is a natural embedding of l^N into $L_N[0,+\infty)$ given by

$$I(\{x_n\}) = x, \text{ where}$$

$$x(t) = x_n \text{ for } n\leq t<n+1,\ n=0,1,2,... \qquad (2.8.20)$$

We show the following

Theorem II.8.7. There holds the following inequality

$$\bar{\Delta}_k(l^N) \leq \bar{\Delta}_k(L_N[0,+\infty)).$$

Proof. Let D be a subspace of codimension k of l^N. The space D can be described by a system $f_1,\dots,f_k$ of linearly independent continuous functionals $f_i \in l^M$, where M is the Young function conjugate to N (see [136]).

i.e. $\qquad D = \{ x \in l^N : f_i(x) = 0, i=1,\dots,k\}.$

Of course, each f_i can be embedded in a natural way into $L_M[0,+\infty)$. This embedding is also given by (2.8.20). Thus we consider f_i as a functional defined on the whole $L_N[0,+\infty)$. Let

$$\hat{D} = \{x \in L_N[0,+\infty) : f_i(x) = 0,\ i=1,\dots,k\} \tag{2.8.21}$$

Of course, $\hat{D}$ is a subspace of codimension k of the space $L_N[0,+\infty)$. By the definition of $\bar{\Delta}_k(L_n[0,+\infty))$, for each $\varepsilon>0$ there is a projection P_o of $L_N[0,+\infty)$ onto $\hat{D}$ such that $\|P_o\| \le \bar{\Delta}_k(L_N[0,+\infty)) + \varepsilon$. The essential problem is that P_o is not necessarily mapping l^N into itself. Thus we shall construct another projection P mapping l^N onto itself with the norm no greater than the norm of P_o. For this purpose we shall introduce a family of isometries T_s, $0<s\le 1$ in $L_N[0,+\infty)$ in the following way:

$$(T_s x)t = \begin{cases} x(t+s) & \text{if } t+s < E(t+s)+1 \\ x(t+s-1) & \text{if } t+s \ge E(t+s)+1, \end{cases}$$

where E(a) denotes the greatest integer less that a. Let

$$P = \int_0^1 (T_s \circ P_o \circ T_s^{-1})\, ds.$$

Since T_s are isometries, we get $\|P\| \le \|P_o\|$. Now we shall show that P is a projection and maps l^N into itself. Recall that P_o can be represented in the form: $P_o x = x - \sum_{i=1}^{k} f_i(x)\cdot z_i$, where z_i are elements with $f_i(z_j) = \delta_{ij}$ (see e.g. [21]) .Since f_i are of the form (2.8.20),

$$f_i(T_s x) = f_i(x) \quad \text{for } i=1,\dots,k,\ 0<s\le 1. \tag{2.8.22}$$

Therefore, by simple calculations, we obtain that

$$Px = x - \sum_{i=1}^{k} f_i(x)\cdot \int_0^1 (T_s z_i)\, ds.$$

Following (2.8.20), $f_i(\int_0^1 (T_s z_j)\, ds) = \delta_{ij}$. Hence P is a projection onto $\hat{D}$.

Observe that for $x \in l^N$, $\int_0^1 (T_s x)\, ds \in l^N$, which implies that $D = P(l^N) \subset \subset l^N$. Since $\|P\| \le \|P_o\|$, the proof of Theorem 8.7 is fully completed.***

Notes and Remarks

1. The concept of a reflection invariant space is due to C. Bessaga. Also formula (2.1.9) is due to him. Example II.1.4 has been first published in [156].

The majority of results presented in the subsequent sections (§2 -§5) have been included in article [155] submitted to the publisher on April 1986.

2. The main results: Theorem II.2.5 and Proposition II.2.2 follow [155]. The method of estimating of the norm of operator $P_{f,z}$ which we employed in Proposition II.2.2 can be without much difficulty carried over to the case where the unit ball of B is a zonotope. A suitable description of the norm $\|P_{f,z}\|$ in the case of B being a reflection invariant space would be of interest.

3. Corollary II.3.9 and formula (2.3.8) in theorem II.3.6 are n-dimensional analogues of the results of J. Blatter and E.W. Cheney concerning the space c_o (see [21], 1974). The main result of this section, Theorem II.3.6, was obtained by the first author in 1979 and published in [155] (see also [156]).

4. Proposition II.4.1 is a direct adaptation of the result of [21] on the norm of a minimal projection in l^1. Lemma II.4.2 has been published in the first author's paper [155]. An independent proof has been found by V.V. Lokot (see [127]). Also Theorem II.4.9 is a routine adaptation of a result of [21]. A straightforward proof, not appealing to [21], has been given in [156]. Example II.4.6, which displays the relevence of minimal projections and their uniqueness to the theory of matrix games, is meant as an illustration.

5. The main result, Theorem II.5.2, was obtained by the first author in December, 1979. A primary version had consisted of several specific cases, which have been later on combined into condition (ii) in its unified form. A simplification is achieved by introducing Tables 1 an 2 of feasible operations; this is owed to C. Bessaga.

In 1983 the first author was informed by V.V. Lokot (oral comunication) about a result of his (joint with his disciples), containing a necessary and sufficient conditions for nonuniqueness of a minimal projection of l_1^n ($n\geq 3$) onto a subspace of codimension one (see [127]). The proof is based on a previous paper of V.V. Lokot [126] (1978) and does not resort the theory of mathematical programming. We now state this result (in notation of the present book).

Theorem II.5.14. (V.V. Lokot and others [127]). let $B = l_1^n$, $n\geq 4$, and let $f \in S_{B^*}$ satisfy $1 \geq f_1 \geq \ldots \geq f_n \geq 0$. Suppose that the norm of a minimal projection onto $f^{-1}(0)$ strictly exceedes one.

A minimal projection onto $f^{-1}(0)$ is nonunique if and only if at least one of the following conditions is satisfied:

1. $f_n > 0$, $\sum_{i=1}^{n} f_i = n - 2$, $\sum_{i=1}^{n} f_i^{-1} > (n-2)\cdot f_n^{-1}$.

2. There exists an integer q with $n > q > 1$ such that

$$\begin{cases} f_i = 1 & \text{for } i \in \{1,\ldots,q\} \\ f_i < 1 & \text{for } i \in \{q+1,\ldots,n\} \end{cases}$$

and the inequalities

$$f_{s+1} > 0,\ (s-1)^{-1}\cdot\sum_{i=1}^{s+1} f_i < 1 \leq (s-2)^{-1}\cdot\sum_{i=1}^{s} f_i,\ \sum_{i=1}^{s+1} f_i^{-1} > (s-1)\cdot f_{s+1}^{-1}$$

hold for a certain $s \in \{3,\ldots,n-1\}$.

3. There exists $s \in \{3,\ldots,n-1\}$ for which the equality $s-2 = \sum_{i=1}^{s} f_i$

and the inequalities

$$0 < f_{s+1} < f_s,\ (s-1)^{-1}\cdot\sum_{i=1}^{s+1} f_i < 1,\ \sum_{i=1}^{s+1} f_i^{-1} \geq (s-1)\cdot f_{s+1}^{-1}$$

hold true.

Let us mention here that condition $(s-1)^{-1}\cdot\sum_{i=1}^{s+1} f_i < 1$ is redundant since, given that $s-2 = \sum_{i=1}^{s} f_i$, it means simply that $f_{s+1} < 1$.

6. Theorem II.6.2 was published in [148] (1980). The proof given here comes from [156]. Propositions II.6.1 and II.6.3 are due to J. Blatter and E.W. Cheney [21]. Theorem II.6.5 follows the paper [153].

7. The results obtained in this section are due to S. Rolewicz [169]. Note that recently M. Baronti and C.Franchetti have calculated the strict value of the constant $\bar{\Delta}_1(l^p)$ (see [11]). They proved the following

Theorem II.7.11. $\bar{\Delta}_1(l^p) = \Delta_1(L^p) = \Lambda_p$, where $\Lambda_p = \max\{\phi_p(t) : t \in [0,1]\}$ and $\phi_p(t) = [(t^{1/(p-1)} + (1-t)^{1/(p-1)}]^{(p-1)/p}\cdot[t^{p-1} + (1-t)^{p-1}]^{1/p}$

8. The results of this section were also established by S. Rolewicz in [169].

Chapter III

Kolmogorov's Type Criteria for Minimal Projections

§ 1. Preliminaries and supplementary notations

Let $C(T,\mathbb{K})$ ($\mathbb{K}=\mathbb{C}$ or $\mathbb{K}=\mathbb{R}$) denote the space of all continuous, $\mathbb{K}$-valued functions defined on a compact set T with the supremum norm $\|\cdot\|$. For $f \in C(T,\mathbb{K})$ and $V \subset C(T,\mathbb{K})$ put

$$P_V(f) = \{v \in V: \|f-v\| = \rho(f,V)\} \tag{3.1.1}$$

If V is a linear subspace of $C(T,\mathbb{K})$ then classical Kolmogorov's criterion reads as follows:

$v \in P_V(f)$ if and only if for every $w \in V$

$$\inf\{re((f(t)-v(t))\cdot\overline{w(t)}): t \in C(f-v)\} \le 0 \tag{3.1.2}$$

where $C(f-v) = \{t \in T: |f(t)-v(t)| = \|f-v\|\}$.

The above characterization of best approximants can be extended to the case of an arbitrary Banach space. Namely, let B be a Banach space over a field $\mathbb{K}$ ($\mathbb{K}=\mathbb{C}$ or $\mathbb{K}=\mathbb{R}$). For $x \in B$ put

$$E(x) = \{f \in \text{ext}\, S_{B^*}: f(x) = \|x\|\}. \tag{3.1.3}$$

and let for $D \subset B$

$$P_D(x) = \{v \in D: \|x-v\| = \rho(x,D)\} \tag{3.1.4}$$

Then there holds the following

Theorem III.1.1. (see [30]) For every $D \subset B$ the following conditions are equivalent:

D is a sun ,i.e. for every $x \in B$, $v \in P_D(x)$ and $t \ge 0$

$$v \in P_D(v+t\cdot(x-v)) \tag{3.1.5}$$

For every $x \in B$; $v \in P_D(x)$ if and only if for every $u \in D$

there exists $f \in E(x-v)$ such that $re(f(u-v)) \le 0$. (3.1.6)

The similar result can be proved in case of strong unicity. In order to present it let us recall that an element $v \in D$ is called a strongly unique best approximation (briefly SUBA) to $x \in B$ if and only if there is $r>0$ such that for every $u \in D$

$$\|x-u\| \ge \|x-v\| + r\cdot\|u-v\| \tag{3.1.7}$$

The theory of strong unicity has its origin in the following result of Newman and Shapiro [137]

Theorem III.1.2. Assume $D \subset C(T,\mathbb{K})$ is a Haar subspace of $C(T,\mathbb{K})$. Then for every $x \in C(T,\mathbb{K})$ there is a constant $r>0$ such that the best approximation $v \in P_D(x)$ satisfies one of the following inequalities:

$$\|x-u\| \geq \|x-v\| + r\cdot\|v-u\| \quad \text{for every } u \in D \text{ in case } \mathbb{K}=\mathbb{R} \tag{3.1.8}$$

and

$$\|x-u\|^2 \geq \|x-v\|^2 + r\cdot\|v-u\|^2 \text{ for every } u \in D \text{ in case } \mathbb{K}=\mathbb{C} \tag{3.1.9}$$

It is obvious that if u is a SUBA to x in D then card $P_D(x) = 1$. The converse is not true (see e.g. [35]). The significance of this notion can be illustrated by E.W. Cheney's observation [35, p.82] that the strong unicity of best approximation yields the continuity of metric projection

$$P_D : X \ni x \to P_D(x) \in D. \tag{3.1.10}$$

(see [184], Th.2.4). Also one can see that the proof of the Remez algorithm depends, in fact, on strong unicity. For more detailed information about strong unicity the reader is referred to [184].

In ([189], Th.2.1, p.885) there was proved the following

Theorem III.1.3. Let $x \in B \setminus D$ and let $D \subset B$ be a starlike set with respect to $v \in D$. Then the following statements are equivalent:

$$v \text{ is a SUBA to } x \text{ in } B \text{ with a constant } r>0 \tag{3.1.11}$$

$$\text{For every } u \in D \ \mathrm{re}(f(u-v)) \leq -r\cdot\|u-v\| \text{ for some } f \in E(x-v). \tag{3.1.12}$$

It is clear that each convex set is a sun and a starlike set with respect to each of its points, so the Theorem 1.1 and 1.3 may be treated as generalizations of Kolmogorov's criterion (3.1.2). However, in general, their applications seem to be limited because in many cases we do not know how the set ext S_{B^*} looks like.

The aim of this chapter is to present, applying Theorems 1.1 and 1.3, various Kolmogorov's type criteria for the case $B = \mathcal{K}(X,Y)$, where $\mathcal{K}(X,Y)$ denotes the space of all compact operators going from a Banach space X to a Banach space Y (not necessary a subspace of X). These characterizations are expressed in terms of the set ext S_{Y^*} which is more convenient for applications. Of course, we concentrate mainly on the case of projections i.e. $B = \mathcal{K}(X,Y)$, $D = \mathcal{P}(X,Y)$, $v = 0$, where $Y \subset X$ is its finite dimensional vector subspace (see sections 3,5,6).

Throughout this chapter, given Banach spaces B and D over the same field $\mathbb{K}$ ($\mathbb{K}=\mathbb{C}$ or $\mathbb{K}=\mathbb{R}$), we write $\mathcal{K}(B,D)$ for the space of all compact operators from B into D. The symbol $\mathcal{L}_e(B^*,D)$ stands for the space of all weak*-weakly continuous compact operators from B^* into D endowed with the operator norm.

For $W \subset B$ we write

$$W^{\circ} = \{f \in B^{*}: |f(x)| \le 1 \text{ for every } x \in W\} \tag{3.1.13}$$

The set W° will be called the polar set to W. Respectively the set

$$W^{\circ\circ} = \{x \in B: \ |f(x)| \le 1 \text{ for every } f \in W^{\circ}\} \tag{3.1.14}$$

will be called the bipolar set to W.
In the sequel we will need the following results

Theorem III.1.4. (see e.g.[2],p.320) Assume B is a locally convex, linear, topological space and let $W \subset B$ be its compact subset. If the set $\mathrm{conv}(\bar{W})$ is compact then

$$\mathrm{ext}\ \mathrm{conv}(W)^{-} \subset W \tag{3.1.15}$$

where $\mathrm{conv}(W)^{-}$ denotes the smallest closed convex set containing W.

Theorem III.1.5. (see e.g.[2],p.339) Let B be a Banach space and let $W \subset B$ be a compact subset. Assume furthermore that W is a ballanced set and that the set $\mathrm{conv}(W)^{-}$ is compact. Then

$$\mathrm{conv}(W)^{-} = W^{\circ\circ} \tag{3.1.16}$$

Proposition III.1.6. (see [51], Ex.(0.2)) The space $\mathcal{K}(B,D)$ is linearly isometric with the space $\mathcal{L}_{e}(B^{**},D)$.

Proof. Fix $T \in \mathcal{K}(B,D)$ and $\phi \in B^{**}$. Following Goldstine's Theorem there exists a net $(x_u) \subset W_B(o,\|\phi\|)$ such that $\hat{x}_u \to \phi$ weak* in B^{**}. Since for each u $\|x_u\| \le \|\phi\|$, the set $\{Tx_u\}_{u\in U}$ is relativly compact. Set for $u \in U$ $C_u = \mathrm{cl}\{Tx_v\}_{v\ge u}$ (the closure is taken with respect to the norm topology in D). Note that $\{C_u\}_{u\in U}$ is a centered family of closed sets. By the compactness of T $\bigcap_{u\in U} C_u \ne \emptyset$. Take $y \in \bigcap_{u\in U} C_u$. We show that Tx_u tends to y weakly in D. To do this fix $\varepsilon>0$ and $f \in D^{*}$. Let $V = \{z \in D: |f(y)| < \varepsilon/2\}$. Since T is continuous with respect to the norm topologies in B and D, T is weakly - weakly continuous. Hence we may select an open neighbourhood W of O (in the weak topology in B) such that $T(W) \subset V$. Since $\hat{x}_u$ tends to ϕ weakly* in B^{**}, $x_v - x_z \in W$ for $z,w \ge u_o$. Choose $w \ge u_o$ with $\|Tx_v - y\| \le \varepsilon/2$ (the existence of such w is guaranteed by the definition of sets C_u. Then for each $z \ge u_o$

$$|f(Tx_z) - y| \le |f(T(x_z - x_v))| + |f(Tx_v - y)| \le \varepsilon/2 + \|f\|\cdot\varepsilon/2,$$

since the first term belongs to V. Consequently, $Tx_u \to y$ weakly in D^{*}. Now we shall show that for every net $(y_v) \subset W_B(O,\|\phi\|)$ tending weak* in B^{**} to ϕ, the net (Ty_v) tends to y weakly in D. Assume this is not true and select a net $(y_v) \subset W_B(0,\|\phi\|)$, $y_v \to \phi$ weak* in B^{**} with $Ty_v \to x \neq y$ weakly in D. Take $f \in S_{D^*}$ with $f(x-y) = \|x-y\|$ and let $V = \{z \in D: |f(z)| < \|x-y\|/2\}$. Reasoning similarly as in the previous part of the proof, we obtain that $|f(Tx_u - Ty_v)| < \|x-y\|/2$ for $u \geq u_o$ and $w \geq w_o$ and consequently $|f(x-y)| < \|x-y\|/2$; contradiction.

Now we are able to define the required isometry. Put for $\phi \in B^{**}$

$$T^*\phi = \lim_u Tx_u, \tag{3.1.17}$$

where $\{x_u\} \subset W_B(0,\|\phi\|)$ is an arbitrary net tending to ϕ weakly* in B^{**} (we have proved that the limit does not depend on the choice of net $\{x_u\}$).

It is clear that T^* is linear and $T^*|_B = T$.

Now we prove that $T^* \in \mathcal{L}_e(B^{**},D)$. Take $\{\phi_u\} \subset B^{**}$, and let ϕ_u tend to 0 weakly* in B^{**}. Fix $\varepsilon>0$, $f \in D^*$ and let $V = \{y \in D: |f(y)| < \varepsilon\}$. Since T is weakly - weakly continuous, there exists an open (in the weak topology of D) neighbourhood W of 0 with $T(W) \subset V$. Note that for $u \geq u_o$ $\phi_u \in W$ (W may be treated as weak* open set in B^{**}). For each $u \geq u_o$ select a net $(x_u^v) \subset B$ such that $\|x_u^v\| \leq \|\phi_u\|$ and (x_u^v) tends weak* in B^{**} to ϕ_u. It is clear that for $w \geq w_u$ $x_u^v \in W$. Consequently, following (3.1.17), $T^*\phi_u = \lim_v Tx_u^v$ belongs to V, which proves the weak*- weak continuity of T^*.

To show that T^* is a compact operator, note that there exists a compact set $K \subset D$ with $T(W_B(0,1)) \subset K$. Following (3.1.17), $T^*(W_{B^{**}}(0,1)) \subset \operatorname{conv}(K)^-$ (see (3.1.15)). By Mazur's Theorem the set $\operatorname{conv}(K)^-$ is compact too. Since $W_D(0,\|T\|)$ is a convex set, by (3.1.17), $T^*(W_{B^{**}}(0,1)) \subset W_D(0,\|T\|)$, which means $\|T^*\| = \|T\|$.

It is clear that for every $S \in \mathcal{L}_e(B^{**},D)$, $S = (S|_B)^*$. Consequently the operator $*$ defined by (3.1.17) is a linear isometry, which completes the proof.***

Remark.III.1.7. If $L \in \mathcal{K}(B,D)$ is a finite dimensional operator then

$$L^*f = \sum_{i=1}^{n} f(x_i^*)\cdot y_i, \text{ for } f \in B^{**}, \tag{3.1.18}$$

where $L = \sum_{i=1}^{n} x_i^*(\cdot)\cdot y_i$.

Before presenting the result which is crucial for our later investigations let us introduce some notations. For $x \in B^*$, $y \in D^*$ and $h \in \mathcal{L}_e(B^*,D)$ put

$(x\otimes y)(h) = y(h(x))$ (3.1.19)

Let us W_B° and W_D° denote the polar set to W_B and W_D (see (3.1.13). Set

$W_B^\circ \otimes W_D^\circ = \{x\otimes y: x \in W_B^\circ,\ y \in W_D^\circ\}$ (3.1.20).

Then there holds the following

Theorem III.1.8. Assume B, D are Banach spaces. Then

$$\text{ext } (W_B^\circ \odot W_D^\circ)^{\circ\circ} \subset \text{ext } W_B^\circ \otimes \text{ext } W_D^\circ. \qquad (3.1.21)$$

(the symbol "$\circ\circ$" denotes the bipolar set to $W_B^\circ \otimes W_D^\circ$ see (3.1.14)).

Proof. At first we show that $W_B^\circ \otimes W_D^\circ$ is a compact set with respect to the weak* topology in $\mathscr{L}_e^*(B^*,D)$. To do this, let us define a map

$$\Phi\colon W_B^\circ \times W_D^\circ \to W_B^\circ \otimes W_D^\circ \text{ by } \Phi(x,y) = x \otimes y \qquad (3.1.22)$$

Following the Banach-Alaoglu Theorem, W_B° (resp. W_D°) is a weak* compact set in B^* (resp. in D^*). Hence it suffices to show that Φ is continuous (in $W_\circ^B \times W_D^\circ$ we consider the Tychonoff topology). Take $\{x_u\} \subset \overset{\circ}{W_B}$, $\{y_u\}\subset W_D^\circ$ with $x_u \to x$ and $y_u \to y$ and fix $h \in \mathscr{L}_e(B^*,D)$. At first we prove that $\|h(x_u-x)\| \to 0$. For the converse suppose that $\|h(x_v-x)\| \geq d > 0$ for some subnet $\{x_v\} \subset \{x_u\}$. Let us set $C_v = \text{cl}\{h(x_z-x)\}_{z\geq v, z\in W}$ for $w \in W$ (the closure is taken with respect to the norm topology in D). By the compactness of h, $\bigcap_{v\in W} C_v \neq \emptyset$. Take $y \in \bigcap_{v\in W} C_v$ and select $f \in S_{D^*}$ with $f(y) = \|y\|$.

For each $n \in \mathbb{N}$ and for every $w \in W$ there exists $z_v \in W$, $z_v \geq w$ with $\|h(x_{z_v}-x)-y\| \leq 1/n$. But it means that $f(h(x_{z_v}-x)) \underset{v}{\to} f(y) = \|y\| \geq d > 0$; contradiction with $h(x_u-x) \to 0$ weakly in D. Now compute

$$|\Phi(x_u,y_u)(h) - \Phi(x,y)(h)| = |(x_u\otimes y_u)h - (x \otimes y)h| \leq$$

$$\leq |y_u(hx_u) - y_u(hx)| + |y_u(hx) - y(hx)| \leq$$

$$\leq \sup\{\|y_u\|: u \in U\}\cdot\|h(x_u-x)\| + |(y_u - y)(hx)|.$$

Since $\sup\{\|y_u\|: u \in U\} < +\infty$, $\|h(x_u-x)\| \underset{u}{\to} 0$ and $y_u \to y$ weakly in D^*, the proof of the continuity of Φ is fully completed. Consequently the set $W_B^\circ \otimes W_D^\circ$ is compact in the weak* topology in $\mathscr{L}_e^*(B^*,D)$. It is easy to show that conv $(W_B^\circ \otimes W_D^\circ)^-$ (recall that conv $(W_B^\circ \otimes W_D^\circ)^-$ is the smallest closed in the weak* topology of $\mathscr{L}_e^*(B^*,D)$, convex set containing $W_B^\circ \otimes W_D^\circ$) is included in $W_{\mathscr{L}_e^*(B^*,D)}$. Following the Banach-Alaoglu Theorem, conv $(W_B^\circ \otimes W_D^\circ)^-$ is a weak* compact set in $\mathscr{L}_e^*(B^*,D)$. Applying Theorem 1.4 we get

$$\text{extr conv}(W^\circ_B \otimes W^\circ_D)^- \subset W^\circ_B \otimes W^\circ_D.$$

By Theorem 1.5, conv $(W^\circ_B \otimes W^\circ_D)^{-\|\cdot\|} = (W^\circ_B \otimes W^\circ_D)^{\circ\circ}$. Folllwing ([51],p.55), $(W^\circ_B \otimes W^\circ_D)^{\circ\circ} = W_{\mathcal{L}^*_\otimes(B^*,D)}$, which gives conv $(W^\circ_B \otimes W^\circ_D)^- =$ conv $(W^\circ_B \otimes W^\circ_D)^{-\|\cdot\|}$. Consequenty ext $(W^\circ_B \otimes W^\circ_D)^{\circ\circ} \subset W^\circ_B \otimes W^\circ_D$ and it implies that

ext $(W^\circ_B \otimes W^\circ_D)^{\circ\circ} \subset$ ext $W^\circ_B \otimes$ ext W°_D, which completes the proof of the theorem. ***

Remark III.1.9. Accordingly to ([51,p.55) $(W^\circ_B \otimes W^\circ_D)^{\circ\circ} = W_{\mathcal{L}^*_\otimes(B^*,D)}$. Hence

$$\text{ext } W_{\mathcal{L}^*_\otimes(B^*,D)} \subset \text{ext } W^\circ_B \otimes \text{ext } W^\circ_D. \tag{3.1.23}$$

Proposition 1.6 and Remark 1.9 yield the following

Corollary III.1.10.

$$\text{ext } W_{\mathcal{K}^*(B,D)} \subset \text{ext } W_{B^{**}} \otimes \text{ext } W_{D^*} \tag{3.1.24}$$

where $(x\otimes y)L = y(L^*x)$ for every $x \in B^{**}$, $y \in D^*$, $L \in \mathcal{K}(B,D)$ (L^* is defined by (3.1.17)).

In sections 4,5 the following notion will be frequently used

Proposition III.1.11. (see [41]) Let $B = C(T,\mathbb{K})$ and let $D \subset C(T,\mathbb{K})$ be its linear subspace. Given $L \in \mathcal{L}(B,D)$ let us set

$$\mathcal{F}_L = \{F\subset T\colon \text{ F is closed and for every } x\in C(T,\mathbb{K})\ Lx = 0 \text{ if } x|_F=0\} \tag{3.1.25}$$

Then there exists the smallest, in the sence of inclusion, set $F_o \in \mathcal{F}$.

Definition III.1.12. Assume that $L \in \mathcal{K}(C(T,\mathbb{K}),D)$. The smallest, in the sence of inclusion, set belonging to $\mathcal{F}_L$ is called the carrier of the operarator L (we write car(L) for brevity).

If the car(L) is is finite, then L is called a discrete operator. The set of all discrete operators going from $C(T,\mathbb{K})$ into D will be denoted by $\mathcal{D}(C(T,\mathbb{K}),D)$ (briefly $\mathcal{D}(B)$ if $B = C(T,\mathbb{K}) = D$). For $F \subset T$ we write $\mathcal{D}(C(T,\mathbb{K}),D,F)$ (briefly $\mathcal{D}(B,F)$ if $B = C(T,\mathbb{K}) = D$) for the set of all $L \in \mathcal{D}(C(T,\mathbb{K}),D)$ with car(L) $\subset$ F. By $\mathcal{P}_D(C(T,\mathbb{K}),D,F)$ we denote the set of all discrete operators $P \in \mathcal{P}(C(T,\mathbb{K}),D)$ such that car(P) $\subset$ F.

Now assume $D \subset C(T,\mathbb{K})$ is its n-dimensional subspace. Then by $I(C(T,\mathbb{K}),D)$ we denote the set of all interpolating projections going from $C(T,\mathbb{K})$ onto D i.e.

$$P \in I(C(T,\mathbb{K}),D) \text{ if and only if } P = \sum_{i=1}^{n} \hat{t}_i(\cdot)\cdot y_i \tag{3.1.26}$$

where $t_i \in T$, $y_i \in D$ for i=1,...,n and $y_i(t_j) = \delta_{ij}$ (i,j=1,...,n).

At the end of this section we present terminology concerning generalised sequence spaces.

Given an arbitrary set T by $c_o(T)$, written c_o for brevity, we denote the space of all functions $x : T \to \mathbb{K}$ such that the set $\{t \in T: |x(t)| > \varepsilon\}$ is finite for all $\varepsilon > 0$. The norm in c_o is $\|x\|_\infty = \sup\{|x(t)|: t \in T\}$. The space $l_1(T)$ consists of all functions $x: T \to \mathbb{K}$ which are zero except on a countable set in T for which $\|x\|_1 = \sum_{t\in T} |x(t)| < +\infty$. It is well known that the conjugate space of c_o can be isometrically identified with $l_1(T)$ (written l_1 for brevity) and the conjugate space of l_1 with l_∞, where

$$l_\infty = \{x: T \to \mathbb{K}: \ \sup\{|x(t)|: t \in T\} < +\infty\}. \tag{3.1.27}$$

We note that

$$\text{ext } S_{l_1} = \{\alpha \cdot f_t : t \in T, \alpha \in \mathbb{K}, |\alpha| = 1\}, \tag{3.1.28}$$

where $f_t(s) = \begin{cases} 0 & ; s \neq t \\ 1 & ; s = t \end{cases}$

and

$$\text{ext } S_{l_\infty} = \{f: T \to \mathbb{K}: |f(t)| = 1 \text{ for every } t \in T\}. \tag{3.1.29}$$

By ([60], Th. 18, p. 274), the set ext $S_{l_\infty^*}$ has the following representation

$$\text{ext } S_{l_\infty^*} = \text{cl}\{\hat{t}: t \in T\}, \tag{3.1.30}$$

where $\hat{t}(f) = f(t)$ for every $f \in l_\infty$ and the closure is taken with respect to the weak* topology in l_∞^*.

§ 2. General case

We start with the following

Lemma III.2.1. Let B and D be a Banach spaces, both over the same field $\mathbb{K}$ ($\mathbb{K} = \mathbb{R}$ or $\mathbb{K} = \mathbb{C}$). For $L \in \mathcal{K}(B,D)$ put

$$\text{crit}^*L = \{f \in \text{ext } S_{D^*}: \|f \circ L\| = \|L\|\} \tag{3.2.1}$$

Then the set $\text{crit}^*(L)$ is nonvoid for every $L \in \mathcal{K}(B,D)$.

Proof. Fix $L \in \mathcal{K}(B,D)$ and consider the function $\phi(f) = \|f \circ L\|$ for $f \in S_{D^*}$. We show that ϕ is weak continuous on S_{D^*}. By the compactness of L the space L(B) is separable. Since $f \circ L = f|_{L(B)} \circ L$, we may restrict ourselves to the case when D is separable. Following ([60], Th. 1, p. 426), the space S_{D^*} with the weak* topology is metrizable in this case.

Now suppose on the contrary that $(f_n) \subset S_{D^*}$ tends weak* to $f \in S_{D^*}$ and $\phi(f_n - f) \geq \varepsilon > 0$. Then $(f_n - f)(Lx_n) > \varepsilon/2$ for some $(x_n) \subset S_B$. By the compactness of L we may assume $\|Lx_n - y\| \to 0$ for some $y \in D$. We note that

$$|(f_n - f)(Lx_n)| \leq |(f_n - f)(Lx_n - y)| + |(f_n - f)(y)| \leq$$

$$\leq 2 \cdot \|Lx_n - y\| + |(f_n - f)(y)| \leq \varepsilon/2 \text{ for } n \geq n_o;$$

contradiction. Applying the Banach-Alaoghlu and the Krein-Milman Theorems we complete the proof. ***

Now we prove the main result of this section.

Theorem III.2.2. Let B,D be such as in Lemma 2.1. Assume $\mathcal{V} \subset \mathcal{K}(B,D)$ is a convex set. Let $K \in \mathcal{K}(B,D)$ and $V \in \mathcal{V}$. Then we have:

(a) $V \in P_{\mathcal{V}}(K)$ (see (3.1.4)) if and only if for every $U \in \mathcal{V}$ there exists $y \in \mathrm{crit}^*(K-V)$ such that $\|\mathrm{re}(y^* \circ (K-U))\| \geq \|K-V\|$.

(b) V is a SUBA to K in $\mathcal{V}$ with a constant $r>0$ if and only if for every $U \in \mathcal{V}$ there exists $y^* \in \mathrm{crit}^*(K-V)$ such that $\|\mathrm{re}(y^* \circ (K-U))\| \geq \|K-V\| + r \cdot \|K-U\|$.

Proof (a). Fix $U \in \mathcal{V}$. Since $\|\mathrm{re}(y^* \circ (K-U))\| \geq \|K-V\|$ for some $y^* \in \mathrm{crit}^*(K-V)$, $V \in P_{\mathcal{V}}(K)$.

To prove the converse, assume that there exists $U \in \mathcal{V}$ such that $\|\mathrm{re}(y^* \circ (K-U)\| < \|K-V\|$ for every $y^* \in \mathrm{crit}(K-V)$. Take an arbitrary $f \in E(K-V)$ (see((3.1.3)). Following Corollary (1.10), $f = x^{**} \otimes y^*$ for some $x^{**} \in \mathrm{ext}\, S^*_{B^{**}}$ and $y^* \in \mathrm{ext}\, S_{D^*}$. Accordingly to Goldstine's Theorem select a net $(x_u) \subset S_B$ such that x_u tends to x^{**} weak* in B^{**}. Since, by (1.17), $\mathrm{re}(y^* \circ (K-V)x_u)$ tends to $\mathrm{re}(y^* \circ (K-V)^* x^{**}) = \mathrm{re}((x^{**} \otimes y^*)(K-V) = \mathrm{re}(f(K-V))$, $y^* \in \mathrm{crit}^*(K-V)$. Hence we have

$$\mathrm{re}(f(U-V)) = \mathrm{re}(f(K-V)) - \mathrm{re}(f(K-U)) = \|K-V\| - \mathrm{re}(y^*((K-U)^* x^{**})) =$$

$$= \|K-V\| - \lim_u \mathrm{re}(y^*((K-U)x_u)) \geq \|K-V\| - \|\mathrm{re}(y^* \circ (K-U))\| > 0.$$

Following Theorem (1.1), $V \notin P_{\mathcal{V}}(K)$; contradiction.

By the same reasoning, applying Theorem (1.3), we can prove part (b) of our theorem. ***

Remark III.2.3. In Theorem 2.2 the set $\mathrm{crit}^*(K-V)$ may be replaced by any set $C \subset \mathrm{crit}^*(K-V)$ such that $\bigcup_{|a|=1} a \cdot C = \mathrm{crit}^*(K-V)$ ($C \cup -C = \mathrm{crit}^*(K-V)$ in the real case) and $a \cdot C \cap b \cdot C = \emptyset$ for $a \neq b$, $|a|=|b|=1$ ($C \cap -C = \emptyset$ in the real case).

Now fix $K \in \mathcal{K}(B,D)$ and for $y^* \in \mathrm{crit}^* K$ put

$$A_{y^*} = \{x \in S_B : y^*(Kx) = \|K\|\} \tag{3.2.2}$$

One can show the following

Remark III.2.4. Let $K \in \mathcal{K}(B,D)$ and $y^* \in \mathrm{crit}^* K$. Then for any $a \in \mathbb{K}$, $|a|=1$

$$a \cdot A_{a \cdot y^*} = A_{y^*} \tag{3.2.3}$$

Now we are able to prove

Theorem III.2.5. Assume B is a reflexive space and let $D, \mathcal{V}, K, V$ be such as in Theorem 2.2. Then we have:

(a) $V \in P_{\mathcal{V}}(K)$ if and only if for every $U \in \mathcal{V}$ there exists $y^* \in \mathrm{crit}^*(K-V)$ such that $\inf\{\mathrm{re}(y^*(U-V)x) : x \in A_{y^*}\} \le 0$.

(b) V is a SUBA to K in $\mathcal{V}$ with a constant $r>0$ if and only if for every $U \in \mathcal{V}$ there exists $y^* \in \mathrm{crit}^*(K-V)$ with $\inf\{\mathrm{re}(y^*(U-V)x) : x \in A_{y^*}\} \le -r \cdot \|U-V\|$.

Proof. (a) Assume $V \notin P_{\mathcal{V}}(K)$. Then $\|K-U\| < \|K-V\|$ for some $U \in \mathcal{V}$. Take an arbitrary $y^* \in \mathrm{crit}^*(K-V)$ and $x \in A_{y^*}$. Compute

$$\mathrm{re}(y^*(U-V)x) = \mathrm{re}(y^*(K-V)x) - \mathrm{re}(y^*(K-U)x) \ge \|K-V\| - \|K-U\| > 0$$

and consequently $\inf\{\mathrm{re}(y^*(U-V)x) : x \in A_{y^*}\} > 0$.

To prove the converse, suppose that $\inf\{\mathrm{re}(y^*(U-V)x) : x \in A_{y^*}\} > 0$ for every $y^* \in \mathrm{crit}^*(K-V)$ (the set A_{y^*} is nonvoid by the reflexivity of D). Take an arbitrary $f \in E(K-V)$. In view of Corollary (1.10) $f = x^{**} \otimes y^*$ for some $y^* \in \mathrm{ext}_{D^*}$ and $x^{**} \in \mathrm{ext}\, S_{B^{**}}$. Since B is a reflexive spacce, $x^{**} = x$ for some $x \in S_B$. It is clear that $y^* \in \mathrm{crit}^*(K-V)$ and $x \in A_{y^*}$. Consequently $\mathrm{re}(f(U-V)) > 0$ and, by Theorem (1.1), $V \notin P_{\mathcal{V}}(K)$.

Applying Theorem (1.3), by the same reasoning, we can prove the part (b) of our theorem. ***

Remark III.2.6. In Theorem 2.5, by Remark 2.4, the set $\mathrm{crit}^*(K-V)$ may be replaced by any set $C \subset \mathrm{crit}^*(K-V)$ satisfying the requirements of Remark 2.3.

Remark III.2.7. If B is an arbitrary Banach space it may occur that the set A_{y^*} is empty. Take for example $B = C_o^{2\pi}$, the space of all real, 2π periodic continuous functions, and let D_n be the space of all trigonometric polynomials of degree $\le n$. Put $\mathcal{V} = \mathcal{P}(B,D_n)$, the space of all projections going from B onto D_n. Following Theorem (0.1.3), the classical Fourier projection F_n is minimal among all projections, which means $F_n \in P_{\mathcal{V}}(0)$. Accordingly to Lemma I.2.7, F_n cannot attain its norm in any point of S_{B^*}. Consequently for every $y^* \in \mathrm{crit}^* F_n$ the set A_{y^*} is empty.

Now we restrict our interest to projections. Following Theorem 2.2 there holds

Theorem III.2.8. Let B be a Banach space and let D be its finite dimensional vector subspace. Assume that $P_o \in \Delta(B,D)$, $\|P_o\| > 1$. Then every set $C \subset \mathrm{crit}^*P$ (we treat $\mathcal{P}(B,D)$ as a subset of $\mathcal{K}(B,B)$) such that

$$\mathrm{crit}^*P_o = \bigcup_{|a|=1} a\cdot C \text{ and } a\cdot C \cap b\cdot C = \emptyset \text{ for } a\neq b, |a|=|b|=1 \qquad (3.2.4)$$

(resp. $C \cup -C = \mathrm{crit}^*P_o$, $C \cap -C = \emptyset$ in the real case)

is linearly dependent over D.

Proof. Take $C \subset \mathrm{crit}^*P_o$ satisfying (3.2.4) and assume that C is linearly independent over D. We can write $C = \{f_1,\ldots,f_k\}$, $k\le n$. If $k<n$ we may add $f_{k+1},\ldots,f_n \in S_{B^*}$ such that $\{f_1|_D,\ldots,f_n|_D\}$ form a basis of D^*. It is clear that the set $\{f_1,\ldots,f_n\}$ is total on D. By ([2],p.74) there exists $y_1,\ldots,y_n \in D$ with $f_i(y_j)=\delta_{ij}$. Define an operator $P:B \to D$ by

$$Px = \sum_{i=1}^{n} f_i(x)\cdot y_i \text{ for } x \in B.$$

It is obvious that $P \in \mathcal{P}(B,D)$. We show that $\|P\| < \|P_o\|$. Following Th.2.2 and Remark 2.3 it suffices to prove that $\|\mathrm{re}(f_i\circ P)\| < \|P_o\|$ for $i=1,\ldots,n$. For any $x \in S_B$ we have $\mathrm{re}(f_i\circ P)(x) = \mathrm{re}(f_i(\sum_{j=1}^{n} f_j(x)\cdot y_j)) = \mathrm{re}(f_i(x)) \le 1$. Hence $\|\mathrm{re}(f_i\circ P)\| \le 1 < \|P_o\|$, which completes the proof.***

Remark III.2.9. If we consider the set $\mathcal{P}(B,D)$ as a subset of $\mathcal{K}(B,D)$ then Theorem 2.7.remains true.

If $\|P_o\| = 1$, then the set C given by (3.2.4) may be dependent or independent over D. Consider the following

Example III.2.10. Let $B = l^n_\infty$ $(n\ge3)$ and let $D = \ker(1/2,1/4,1/4,0,\ldots,0)$. Take $y = (2,0,\ldots,0)$ and define $P_y x = x - f(x)\cdot y$ for $x \in B$. Following Proposition II.3.1, it is easy to verify that $\|P_y\| = 1$. We note that for every $i \in \{1,\ldots,n\}$

$$\|e_i\circ P_y\| = \sup\{|x_i-(\sum_{j=1}^{n} f_j\cdot x_j)\cdot y_i| : x \in S_B\} = \qquad (3.2.5)$$

$$= |1-f_i\cdot y_i| + |y_i|\cdot(1-|f_i|).$$

Hence $\mathrm{crit}^*P_y = \{\pm e_i\}_{i=1}^n$. If we put $C = \{e_i\}_{i=1}^n$ then C is linearly dependent on D since $e_1|_D = (-1/2)\cdot e_2|_D + (-1/2)\cdot e_2|_D$.

However if we take $D = \ker(2/3,1/3,0,\dots,0)$ then, by the same reasoning, $\mathrm{crit}^*P_y = \{\pm e_i\}_{i=2}^n$ (we take $y = (3/2,0,\dots,0)$). It is easy to verify that $C = \{e_i\}_{i=2}^n$ satisfies (3.2.4) and it is linearly independent on D.

Now we consider the case $B = C(T,\mathbb{K})$.

Corollary III.2.11. Let $B = C(T,\mathbb{K})$ and let $D \subset B$ be its n-dimensional subspace. If $P_o \in \Delta(B,D)$, $\|P_o\| > 1$, then the set $\{t \in T: \|\hat{t}\circ P_o\| = \|P_o\|\}$ is linearly dependent on D. In particular, if D is a Haar subspace, then $\mathrm{card}\{t \in T: \|\hat{t}\circ P_o\| = \|P_o\|\} \geq n+1$.

Now we restrict our attention to the case $T = [a,b]$ and $D = P_n$ (the space of all polynomials of degree $\leq n$ restricted to $[a,b]$). Following ([101]), if $P \in I(C([a,b],\mathbb{R}),D)$ (see (3.1.26)) then the set $\{t \in [a,b]: \|\hat{t}\circ P\| = \|P\|\}$ consists of at most n points. Note that, by [41] if $D \subset C([a,b],\mathbb{R})$ is an n-dimensional Haar subspace containing constants ($n\geq 3$) $\rho(C([a,b],\mathbb{R}),D) > 1$. Hence, following Corollary 2.11, we get

Corollary III.2.12. In the set $I(C([a,b],\mathbb{R}),P_n)$ ($n\geq 3$) there is no minimal projection going from $C([a,b],\mathbb{R})$ onto P_n.

At the end of this section, applying Theorem 2.8, we give another proof of Theorem II.3.6. Note that, following Proposition II.3.1, we may consider the case $f>0$ and by Proposition II.2.8 we may restrict ourselves to projections P_y induced by $y \in l_\infty^n$, $y \geq 0$. So assume $P_o \in \mathcal{P}(l_\infty^n, \ker f)$ is a minimal projection. Hence, following Theorem 2.8, each set $C \subset \mathrm{crit}^*P_o$ satisfying (3.2.4) is linearly dependent over $\ker(f)$. It is easy to verify that each set $C \subseteq \{e_i\}_{i=1}^n$ is total over $\ker(f)$, since $f > 0$, and consequently it is linearly independent over $\ker(f)$. Hence $\mathrm{crit}^*P_o = \{e_i\}_{i=1}^n \cup \{-e_i\}_{i=1}^n$. Since the corresponding to P_o vector y^o is nonnegative, by (3.2.5), the numbers (y_i^o) ($i=1,\dots,n$), $\|P_o\|$ satisfy the system of equations

$$1 + y_i^o\cdot(1-2\cdot f_i) = \|P_o\| \quad \text{for } i=1,\dots,n$$

$$\sum_{i=1}^n f_i\cdot y_i^o = 1 \qquad (3.2.6)$$

Since the solution of this system is unique , we get after simple calculations

$$\|P_o\| = \Big(\sum_{i=1}^n f_i/(1-2f_i)\Big)^{-1} + 1$$

$$y_i^o = (\|P_o\| - 1)/(1-2f_i) \quad \text{for } i=1,\dots,n \tag{3.2.7}$$

which gives another proof of Theorem II.3.6.

§ 3. SUBA projections onto hyperplanes in l_∞^n and l_1^n

The problems considered in this section may be treated as a developement of results obtained in sections II.5 and II.6. Hence we restrict ourselves to the case $B = l_\infty^n$ (resp. $B = l_1^n$), $\mathcal{V} = \mathcal{P}(B,D) \subset \mathcal{K}(B,B)$, where $D \subset B$ is a hyperplane in l_∞^n (resp. in l_1^n). In other words we will examine in which cases a projection $P_o \in \Delta(B,D)$ is a SUBA to 0 (see(3.1.7) in $\mathcal{P}(B,D)$ i.e. when a projection $P_o \in \Delta(B,D)$ satisfies the inequality:

$$\|P\| \geq \|P_o\| + r\cdot\|P-P_o\| \tag{3.3.1}$$

for every $P \in \mathcal{P}(B,D)$ with constant $r>0$ independent of P.

At first we deal with the case $B = l_\infty^n$. Applying Theorems III.2.2 and III.2.5 we prove the following

Theorem III.3.1. Let $D \subset l_\infty^n$ be a hyperplane i.e. $D = \ker f$ for some $f = (f_1,\dots,f_n) \in l_1^n$, $\|f\| = 1$. Assume $P_o \in \mathcal{P}(l_\infty^n,D)$ is a minimal projection. Then we have:

(a) If $\|P_o\| = 1$, then P_o satisfies (3.3.1) if and only if $|f_i| \geq 1/2$ for exactly one index $i \in \{1,\dots,n\}$.
The constant $r = \min\{1-2\cdot|f_j|; j\neq i\}$ is the best possible.

(b) In the real case, if $\|P_o\| > 1$ then P_o satisfies (3.3.1) if and only if $0 < |f_i| < 1/2$ for $i=1,\dots,n$.
Moreover, the constant

$$r = \min\{\max\{(1-2\cdot|f_i|)\cdot y_i : i=1,\dots,n\} : y \in S_D\} \tag{3.3.2}$$

is the best possible and there holds an estimation

$$r \geq (1-2\cdot|f_j|)\cdot|f_i|/(1-|f_i|), \tag{3.3.3}$$

where $|f_j| = \max\{|f_k| : k=1,\dots,n\}$ and $|f_i| = \min\{|f_k| : k=1,\dots,n\}$.

Proof.(a). Assume that $|f_i| \geq 1/2$ for exactly one index $i \in \{1,\dots,n\}$. Let $P \in \mathcal{P}(B,D)$ be fixed. Denote by y^P (resp. y^o) the corresponding to P (resp. to P_o) vector from B. It is clear that $P - P_o = f(\cdot)\cdot(y^o - y^P)$ and consequently $\|P - P_o\| = \|y^o - y^P\|_\infty$. Since $|f_i| \geq 1/2$, $\|y^o - y^P\|_\infty = |y^P_j - y^o_j|$ for some $j\neq i$. By ([21], Cor. 1), $y^o_i = 1/f_i$ and $y^o_j = 0$ for $j\neq i$. Consequently $\|P - P_o\| = |y^P_j|$ for some $j\neq i$. Following (3.2.5), we note that

$$\|P\| \geq \|(x\to x_j)\circ P\| = |1-f_j\cdot y^P| + |y^P_j|\cdot(1-|f_j|) \geq 1 + |y^P_j|\cdot(1-2\cdot|f_j|) \geq$$
$$\geq \|P_o\| + \min(1-2\cdot|f_k|: k\neq i)\cdot\|P-P_o\|,$$

which gives the result.

Now we shall show that the constant $r = \min(1-2\cdot|f_j|: j\neq i)$ is the best possible. Since $\|P_{f,y}\| = \|P_{|f|,\bar{y}}\|$ for every $f \in l^n_1$ and $y \in \ker(f)$, ($\bar{y}_i = y_i$ if $f_i=0$ and $\bar{y}_i = f_i/|f_i|\cdot y_i$ in the other case) we may assume $f\geq 0$. Set $y_k=0$, if $k\neq i$ and $k\neq j$, $y_i = -f_j/f_i$, $y_j=1$ and let $y = (y_1,\dots,y_n)$ (the index j is so chosen that $f_j = \max(f_k: k\neq i)$). Let $P = P_o - f(\cdot)\cdot y$. By Theorem III.2.2 and Remark III.2.3, it is enough to show that

$$\|(x\to x_k)\circ P\| < 1 + r_1\cdot\|P-P_o\| \text{ for every } r_1 > r \text{ and } k=1,\dots,n.$$

At first we note that $\|P-P_o\| = \|y\|_\infty = 1$. Following (3.2.5), if $k=i$ then

$$\|(x\to x_k)\circ P\| = |1-f_i\cdot(y_i + 1/f_i)| + |y_i + 1/f_i|\cdot|1-f_i| =$$
$$= 1/f_i - 1 + y_i\cdot(1-2\cdot f_i) = 1/f_i - 1 + f_j\cdot(2\cdot f_i - 1)/f_i \leq$$
$$\leq 1/f_i - 1 + (1-f_i)\cdot(2\cdot f_i - 1)/f_i = 2\cdot(1-f_i) \leq 1 < 1 + r_1\cdot\|P-P_o\|.$$

If $k\neq i$ and $k\neq j$, then $y^P_k = y_k = 0$. Hence

$$\|(x\to x_k)\circ P\| = 1 < 1 + r_1\cdot\|P-P_o\|.$$

If $k=j$, then

$$\|(x\to x_k)\circ P\| = 2 - 2\cdot f_j = 1 + r\cdot\|P-P_o\| < 1 + r_1\cdot\|P-P_o\|.$$

Applying Theorem (III.2.2 (b)), we complete the proof of part (a).

(b). As in the previous case we may assume $f_i \geq 0$ for $i=1,\dots,n$. Let us define a function $\phi: S_D \to \mathbb{R}$ by the formula

$$\phi(y) = \min((2\cdot f_i - 1)\cdot y_i : i=1,\dots,n). \qquad (3.3.4)$$

Since $f_i > 0$ for $i=1,\dots,n$, $\phi(y) < 0$ for every $y \in S_D$. Hence, by the argument of compactness and continuity of ϕ, the constant $\gamma = \max\{\phi(y): y \in S_D\}$ is negative. We show that P_o is a SUBA to 0 in $\mathcal{P}(l^n_\infty, D)$ with $r = -\gamma$. To do this, following Theorem III.2.5, Remark III.2.6 and Theorem II.3.6, it is enough to prove that for every $P \in \mathcal{P}(l^n_\infty, D)$ there exists $i \in \{1,\dots,n\}$ with

$$\inf\{((P-P_o)x)_i : x \in A_i(P_o)\} \leq -r\cdot\|P-P_o\| \qquad (3.3.5)$$

(we write $A_i(P_o)$ instead of $A_{x\to x_i}(P_o)$ (see 3.2.2))

It is clear that $\|P-P_o\| = \|y^P - y^o\|_\infty$. Set $y = (y^P - y^o)/\|y^P - y^o\|_\infty$ (if $y^P = y^o$

the inequality (3.3.1) is satisfied). Select $i \in \{1,\dots,n\}$ with $\phi(y) = (2\cdot f_i - 1)\cdot y_i$. Following (3.2.5) and (3.2.7), $x \in A_i(P_o)$ iff

$x_j = -\operatorname{sgn}(f_j) = -1$ for $j \neq i$ and $x_i = \operatorname{sgn}(1-f_i\cdot y_i^o) = 1$.

Hence for $x \in A_i(P_o)$ we get

$$((P-P_o)x)_i = f(x)\cdot\|y^P - y^o\|_\infty\cdot y_i = (2\cdot f_i - 1)\cdot y_i\cdot\|y^P - y^o\|_\infty \leq$$
$$\leq -r\cdot\|y^P - y^o\|_\infty.$$

Following Remark III.2.6, we have proved our claiming.

Now we will show that $r \geq (1-2\cdot f_j)\cdot f_i/(1-f_i)$,

where $f_j = \max\{f_k : k=1,\dots,n\}$ and $f_i = \min\{f_k : k=1,\dots,n\}$. To do this, take $y \in S_D$. If $y_k = 1$ for some $k \in \{1,\dots,n\}$, then

$$\phi(y) \leq 2\cdot f_k - 1 \leq 2\cdot f_j - 1 \leq (2\cdot f_j - 1)\cdot(f_i/(1-f_i)),$$

since $f_j < 1/2$ and $f_i < 1/2$.

In the opposite case $y_k = -1$ for some $k \in \{1,\dots,n\}$ and an easy calculation shows that $y_l \geq f_i/(1-f_i)$ for some $l \in \{1,\dots,n\}$. We note that

$$\phi(y) \leq (2\cdot f_l - 1)\cdot y_l \leq (2\cdot f_l - 1)\cdot f_i/(1-f_i) \leq (2\cdot f_j - 1)\cdot f_i/(1-f_i),$$

since $f_l < 1/2$ and $f_j \geq f_l$.

Hence $\gamma \leq (2\cdot f_j - 1)\cdot f_i/(1-f_i)$ and consequently $r \geq (1 - 2\cdot f_j)\cdot f_i/(1-f_i)$.

To prove that the constant r is the best possible,take $r_1 > r$, choose $y \in S_D$ with $\phi(y) > -r_1$, and define $P \in \mathcal{P}(l^n_\infty, D)$ by $P = P_o + f(\cdot)\cdot y$. For $l \in \{1,\dots,n\}$ and $x \in A_l(P_o)$ we have $((P - P_o)x)_l = f(x)\cdot y_l = (2\cdot f_l - 1)\cdot y_l \geq \phi(y) > -r_1 = -r_1\cdot\|P - P_o\|$.

Following Theorem III.2.5 and Remark III.2.6, the proof of the part (b) is fully completed.***

Remark III.3.2. In the complex case Theorem 3.1 (b) does not hold.

Proof. As in the proof of Theorem 3.1 (b) we may assume $f \geq 0$. It is easy to show that the projection P_o considered in Theorem 2.1 (b) is also minimal in the complex case. Following Remark III.2.4

$A_i(P_o) = \alpha\cdot A_{x \to \alpha\cdot x_i}(P_o)$ for every $\alpha \in \mathbb{C}$, $|\alpha|=1$. Hence we may restrict ourselves to the case $\alpha = 1$.

Take $w \in \mathbb{R}^n \cap S_D$ and let $y = 0 + i\cdot w$. For $L = f(\cdot)\cdot y$, $j=1,\dots,n$ and $x \in A_j(P_o)$ we have $re(Lx)_j = re(f(x)\cdot y_j) = (2\cdot f_j - 1)\cdot re(y_j) = 0 > -r\cdot\|y\|$ for every $r > 0$.

Hence, by Theorem III.2.5 (b) and Remark III.2.6, P_o does not satisfy (3.3.1) with any constant $r>0$.***

However, adopting the reasoning from Theorem II.2.6, we can show that the conditions given in Theorem 3.1 (b) are equivalent to the uniqueness of minimal projection in the complex case.

By similar reasoning to that of Theorem 3.1 (a) and Proposition II.6.1 we get

Corollary III.3.3. Let $D \subseteq c_o$ be a hyperplane, $D = \ker(f)$ for some $f \in S_{l_1}$ (we consider the real and complex case). Then the projection $P_o \in \Delta(c_o, D)$ is strongly unique if and only if $|f_i| \geq 1/2$ for exactly one index i.

Now we consider more difficult case $B = l_1^n$.

Proposition III.3.4. Assume $f = (f_1, \dots, f_n) \in S_B$ and let P_o $\|P_o\| = 1$. Then P_o is a unique minimal projection if and only if P_o is a SUBA to 0 in $\mathcal{P}(B, \ker(f))$.

Proof. In view of Proposition II.1.6 we may assume $1 = f_1 \geq f_2 \geq \dots \geq f_n \geq 0$. Following Theorem II.5.1 card $\Delta(B, \ker(f)) = 1$ and $\rho(B, \ker(f)) = 1$ f and only if $f_2 > 0$ and $f_3 = f_4 = \dots = f_n = 0$. So assume $1 = f_1 \geq f_2 > 0 = f_3 = \dots = f_n$. It is easy to verify that if we put $y = (1/(1+f_2), 1/(1+f_2), 0, \dots, 0)$ then the operator

$$P_y x = x - f(x)\cdot y \quad (x \in B) \tag{3.3.6}$$

belongs to $\Delta(B, \ker(f))$. We show that P_y is a SUBA to 0 in $\mathcal{P}(B, \ker(f))$ with the constant $r = f_2$. So take an arbitrary $P \in \mathcal{P}(B, \ker(f))$ and write P in the form $P = I - f(\cdot)\cdot y^P$. Since $\|f\|_\infty = 1$, $\|P - P_o\| = \|y_P - y\|$.

If $y_P^1 < 0$ then $\|y^P - y\| = \|(y_1^P - 1/(1+f_2), y_2^P - 1/(1+f_2), y_3^P, \dots, y_n^P)\| = = \sum_{i=1}^n |y_i^P| = \|y^P\|$. Hence, by Proposition II.4.1,

$\|P\| \geq \|Pe_1\| = |1 - y_1^P| + \|y^P\| - |y_1^P| = 1 - y_1^P + \|y^P\| + y_1^P = 1 + \|y_P\| =$
$= 1 + \|P_y - P\| \geq \|P_y\| + f_2 \cdot \|P - P_y\|$.

If $y_2^P < 0$, by the same reasoning we have $\|P\| \geq \|P_y\| + f_2 \cdot \|P_y - P\|$.

Now suppose that $y_1^P > y_2^P > 0$. It is easy to verify that in this case $\|y - y^P\| = \|y^P\| - 2y_2^P$, since $y_1^P + f_2 \cdot y_2^P = 1$. Compute

$$\|P\| \geq \|Pe_2\| = |1 - f_2 \cdot y_2^P| + f_2 \cdot (\|y^P\| - y_2^P) =$$
$$= 1 + f_2 \cdot (\|y^P\| - 2y_2^P) = \|P_y\| + f_2 \cdot \|y - y^P\| = \|P_y\| + \|P - P_y\| \tag{3.3.7}$$

If $y_2^P > y_1^P > 0$ we have $\|y^P - y\| = \|y^P\| - 2y_1^P$. Hence

$$\|P\| \geq \|Pe_1\| = 1 - y_1^P + \|y^P\| - y_1^P \geq \|P_y\| + f_2 \cdot \|P - P_y\|.$$

Since the strong unicity of P_y implies that P_y is a unique minimal projection, the proof of Proposition III.3.4 is completed.***

Remark III.3.5. Since Proposition II.4.1 holds true in the complex case, Proposition III.3.4 is valid in the complex case too.

Remark III.3.6. The constant f_2 in Proposition 3.4 is the best possible.

Proof. Let $y^P = (y_1^P, y_2^P, 0, \ldots, 0)$, $y_1^P > y_2^P > 0$ and $y_1^P + f_2 \cdot y_2^P = 1$. Since $\|Pe_i\| = 1$ for $i>2$ and $\|Pe_1\| = 1 + \|y^P\| - 2y_1^P < 1$, $\|P\| = \|Pe_2\|$. Accordingly to (3.3.7), $\|P\| = \|P_y\| + f_2 \cdot \|P_y - P\|$, which proves our claim.***

Now we shall investigate more difficult case, when a norm of minimal projection is greater that one. Following Proposition II.1.6 and Proposition II.4.1, in the sequel we may assume that $f = (f_1, \ldots, f_n)$, $n \geq 3$ and $1 = f_1 \geq f_2 \geq \ldots \geq f_n$, $f_3 > 0$. At first we prove some preliminary results.

Lemma III.3.7. Let $f \in S_B$ and let $f_n > 0$. Set for $m \in \{3, \ldots, n\}$ $a_m = \sum_{j=1}^{m} f_j$. If $a_m > m-2$ and there exists $y \in \ker(f) \setminus \{0\}$ satisfying the system of inequalities

$$y_j \geq \sum_{\substack{i=1 \\ i \neq j}}^{m} y_i + \sum_{i=1}^{n-m} |y_{i+m}| \text{ for } j=1, \ldots, m. \tag{3.3.8}$$

Then we can find $y^1 \in \ker(f) \setminus \{0\}$ with

$$y_j^1 > \sum_{\substack{i=1 \\ i \neq j}}^{m} y_i^1 + \sum_{i=1}^{n-m} |y_{i+m}^1| \text{ for } j=1, \ldots, m \tag{3.3.9}$$

$\left(\sum_{i=1}^{n-n} |y_{i+n}| = 0 \text{ by definition} \right)$.

Proof. Take $y \in \ker(f) \setminus \{0\}$ satisfying system (3.3.7) and consider two cases.

Case I. There exists $j \in \{1, \ldots, m\}$ with $y_j > \sum_{\substack{i=1 \\ i \neq j}}^{m} y_i + \sum_{i=1}^{n-m} |y_{i+m}|$. Then we can find $\vartheta > 0$ such that

$$y_j - \vartheta > \sum_{\substack{i=1 \\ i \neq j}}^{m} y_i + \sum_{i=1}^{n-m} |y_{i+m}| + (m-1) \cdot \vartheta \cdot f_j / (a_m - f_j).$$

Define $y_j^1 = y_j - \vartheta$, $y_i^1 = y_i + \vartheta \cdot f_j / (a_m - f_j)$ for $i \in \{1, \ldots, m\} \setminus \{j\}$, $y_i^1 = y_i$ for $i = m+1, \ldots, n$ and put $y^1 = (y_1^1, \ldots, y_n^1)$. Note that

$$\sum_{i=1}^{n} f_i \cdot y_i^1 = \sum_{i=1}^{m} f_i \cdot y_i^1 + \sum_{i=m+1}^{n} f_i \cdot y_i = f_j \cdot y_j - f_j \cdot \vartheta + \sum_{\substack{i=1 \\ i \neq j}}^{m} f_i (y_i + f_j \cdot \vartheta / (a_m - f_j)) +$$

$$+ \sum_{i=m+1}^{n} f_i \cdot y_i = \sum_{i=1}^{n} f_i \cdot y_i = 0.$$

To finish the proof, fix $i \in \{1,\dots,m\}$, $i\neq j$. Since $a_m > m-2$, $a_m - f_j > f_j\cdot(m-3)$, which gives

$$f_j\cdot\vartheta/(a_m-f_j) > (m-2)\cdot f_j\cdot\vartheta/(a_m-f_j)-\vartheta. \tag{3.3.10}$$

Adding (3.3.8) to (3.3.10) we obtain $\displaystyle y^1_i > \sum_{\substack{k=1\\k\neq i}}^{m} y^1_k + \sum_{k=1}^{n-m}|y^1_{k+m}|$

which established formula (3.3.9).

Case II. For every $j \in \{1,\dots,m\}$

$$y_j = \sum_{\substack{i=1\\i\neq j}}^{m} y_i + \sum_{i=1}^{n-m}|y_{i+m}|. \tag{3.3.11}$$

Hence for each $j \in \{1,\dots,m\}$ $y_1 = y_2 = \dots = y_m$ (To show it we subtract for fixed $j,u \in \{1,\dots,m\}$ equalities (3.3.11)). Consequently, by (3.3.11)

$$-(m-2)\cdot y_1 = \sum_{i=1}^{n-m}|y_{i+m}|. \tag{3.3.12}$$

Hence if $n=m$, $y=0$; contradiction. In the opposite case compute

$$0 = \sum_{i=1}^{n} f_i\cdot y_i = \sum_{i=1}^{m} f_i\cdot y_i + \sum_{i=1}^{n-m} f_{i+m}\cdot y_{i+m} = y_1\cdot(\sum_{i=1}^{m} f_i) + \sum_{i=1}^{n-m} f_{i+m}\cdot y_{i+m} \le$$

$$\le y_1\cdot a_m + \sum_{i=1}^{n-m}|f_{i+m}|\cdot|y_{i+m}| \le y_1\cdot a_m + \sum_{i=1}^{n-m}|y_{i+m}| < y_1\cdot(m-2) + \sum_{i=1}^{n-m}|y_{i+m}|=0,$$

since accordingly to (3.3.12) $y_1 < 0$ and $a_m > m-2$. So we may exclude case II. The lemma is proved.***

Lemma III.3.8. Let $f \in S_B$, $f = (f_1,\dots,f_n)$, $n\ge 3$, $f_3>0$, $f_2<1$. Let $m \in \{3,\dots,n\}$ satisfies $a_m < m-2$, $a_{m-1} > m-3$. If there exists $y \in \ker(f) - \{0\}$ satisfying the system of inequalities

$$y_j \ge \sum_{\substack{i=1\\i\neq j}}^{m-1} y_i + \sum_{i=1}^{n-m+1}|y_{i+m-1}| \text{ for } j=2,\dots,m-1$$

$$y_m \ge \sum_{i=1}^{m-1} y_i + \sum_{i=1}^{n-m}|y_{i+m}|. \tag{3.3.13}$$

Then there exists $y^1 \in \ker(f) \setminus \{0\}$ with

$$y^1_j > \sum_{\substack{i=1\\i\neq j}}^{m-1} y^1_i + \sum_{i=1}^{n-m+1}|y^1_{i+m-1}| \text{ for } j=2,\dots,m-1$$

$$y^1_m > \sum_{i=1}^{m-1} y^1_i + \sum_{i=1}^{n-m}|y^1_{i+m}|. \tag{3.3.14}$$

Proof. Take $y \in \ker(f) \setminus \{0\}$ satisfying (3.3.13) and consider three cases.

Case I. $y_m > \sum_{i=1}^{m-1} y_i + \sum_{i=1}^{n-m} |y_{i+m}|.$

Then we can select $\vartheta>0$ with $y_m > \sum_{i=1}^{m-1} y_i + \sum_{i=1}^{n-m} |y_{i+m}| - \vartheta + (m-2)\cdot\vartheta/(a_{m-1}-1).$

Define $y_1^1 = y_1 - \vartheta$, $y_j^1 = y_j + \vartheta/(a_{m-1}-1)$ $(j=2,\dots,m-1)$, $y_j^1 = y_j$ for $j=m,\dots,n$ and set $y^1 = (y_1^1,\dots,y_n^1)$. Compute

$$f(y^1) = \sum_{i=1}^{n} f_i\cdot y_i^1 = y_1 - \vartheta + \sum_{i=2}^{m-1} f_i\cdot(y_i + \vartheta/(a_{m-1}-1)) + \sum_{i=m}^{n} f_i\cdot y_i = \sum_{i=1}^{n} f_i\cdot y_i = 0.$$

Since $a_{m-1} > m-3$, $\vartheta/(a_{m-1}-1) > -\vartheta + (m-3)\cdot\vartheta/(a_{m-1}-3)$. Combining this inequality with (3.3.13), we get

$$y_j^1 > \sum_{\substack{i=1\\ i\neq j}}^{m-1} y_i^1 + \sum_{i=1}^{n-m+1} |y_{i+m-1}^1| \text{ for } j=2,\dots,m-1$$

$$y_m^1 > \sum_{i=1}^{m-1} y_i^1 + \sum_{i=1}^{n-m} |y_{i+m}^1|, \text{ which proves our claim.}$$

Case II. There exists $j \in \{2,\dots,m-1\}$ with $y_j > \sum_{\substack{i=1\\ i\neq j}}^{m-1} y_i + \sum_{i=1}^{n-m+1} |y_{i+m-1}|.$

Hence $y_j - f_j^{-1}\cdot\vartheta > \sum_{\substack{i=1\\ i\neq j}}^{m-1} y_i + \sum_{i=1}^{n-m+1} |y_{i+m-1}| + \vartheta$ for $\vartheta>0$ sufficiently small.

Since $f_2 < 1$,

$$\vartheta - f_j^{-1}\cdot\vartheta < 0. \tag{3.3.15}$$

Define $y^1 = (y_1^1,\dots,y_n^1)$, where $y_1^1 = y_1 + \vartheta$, $y_j^1 = y_j - f_j^{-1}\cdot\vartheta$, $y_i^1 = y_i$ for $i\neq 1,j$.

It is clear that $y^1 \in \ker(f)$. Adding (3.3.14) to (3.3.15) we get for each $k \in \{2,\dots,m-1\}\setminus\{j\}$

$$y_k^1 > \sum_{\substack{i=1\\ i\neq k}}^{m-1} y_i^1 + \sum_{i=1}^{n-m+1} |y_{i+m-1}^1|$$

and $y_m^1 > \sum_{i=1}^{m-1} y_i^1 + \sum_{i=1}^{n-m} |y_{i+m}^1|$, which completes the proof of this case.

Case III.

$$y_j = \sum_{\substack{i=1\\ i\neq j}}^{m-1} y_i + \sum_{i=1}^{n-m+1} |y_{i+m-1}| \text{ for } j=2,\dots,m-1 \tag{3.3.16}$$

and

$$y_m = \sum_{i=1}^{m-1} y_i + \sum_{i=1}^{n-m} |y_{i+m}|. \tag{3.3.17}$$

At first we show that $y_m>0$. Assume, it is not true. Hence, by (3.3.16), for every $j \in \{2,\dots,m-1\}$

$$y_j = \sum_{\substack{i=1\\ i\neq j}}^{m-1} y_i + \sum_{i=2}^{n-m+1} |y_{i+m-1}| - y_m \tag{3.3.18}$$

Subtracting equalities (3.3.16) for fixed $j,k \in \{2,\dots,m-1\}$ we get $y_2=y_3=,\dots,=y_{m-1}$. Following (3.3.17) and (3.3.18) we get $y_{m-1}=0$, which gives $0 = \sum_{i=1}^{n-m+1} |y_{i+m-1}| + y_1$. Since $f_2 < 1$ and $y \in \ker(f)$, $y_{i+m-1}= 0$ for $i=1,\dots,n-m+1$ and consequently $y = 0$: contradiction.

Hence $y_m > 0$ and reasoning as above we get $y_2=y_3=,\dots,=y_{m-1}$. Subtracting (3.3.16) from (3.3.17) we get $y_2 - y_m = y_m - y_2$, which gives $y_2=y_3=,\dots,=y_m > 0$. Following (3.3.17), $y_1 = -(m-3)\cdot y_m - \sum_{i=1}^{n-m} |y_{i+m}|$.

Compute

$$0 = \sum_{i=1}^{n} f_i\cdot y_i = \sum_{i=1}^{m} f_i\cdot y_i + \sum_{i=m+1}^{n} f_i\cdot y_i = -(m-3)\cdot y_m - \sum_{i=m+1}^{n} |y_i| +$$
$$+ (a_m - 1)\cdot y_m + \sum_{i=m+1}^{n} f_i\cdot y_i = (a_m-(m-2))\cdot y_m + \sum_{i=m+1}^{n} f_i\cdot y_i - \sum_{i=m+1}^{n} |y_i| < 0,$$

since $a_m < m-2$ and $y_m>0$.

Thus we can exclude case III and the proof of lemma is fully completed. ***

Remark III.3.9. Assume $P \in \mathcal{P}(B,D)$, $D = \ker(f), f =(1,f_2,\dots,f_n), n\geq 3$ and $f_3 > 0$. Put $C_i= \{g \in \text{ext}\, S_B*: \pm g(Pe_i) = \|Pe_i\|\}$ for $i=1,\dots,n.$. Then $g \in \text{crit}^*P$ if and only if $g \in \bigcup_{i\in A} C_i$, where

$$A = \{i \in \{1,\dots,n\}: \|Pe_i\| = \|P\|\}. \tag{3.3.19}$$

Proof. Assume $g \in \text{crit}^*P$. Since $\text{ext}\, S_B* = \{\pm e_i\}_{i=1}^{n}$, we have $\pm(g\circ P)e_i = \|P\|$ for some $i \in \{1,\dots,n\}$. It is clear that $i \in A$. The converse is obvious. ***

Lemma III.3.10. Let $f \in S_B*$ and let $m = m(f)$ be so chosen as in Theorem II.5.2. Assume that $P_o \in \mathcal{P}(B,\ker(f))$ is a unique minimal projection. Then

a) If $a_m > m-2$ and $m = k(f)$ (see Theorem II.4.9) then $A = \{1,\dots,m\}$.

If $a_m > m-2$ and $m < k(f)$ then $A = \{1,\dots,l\}$, where $l = \max\{i\geq m+1: f_i^{-1} = \beta_m\}$.

b) If $f_2 < 1$, $a_{m-1} > m-3$ and $a_m < m-2$, then $A = \{2,\dots,l\}$, where $l = \max\{i\geq m: f_i=f_m\}$.

Proof. a). Following (2.4.20) and Proposition II.5.7, corresponding to P_o vector y^o has coordinates $y^o_1 = u\cdot(\beta_m - f_1^{-1})/2, \dots, y^o_m = u\cdot(\beta_m - f_m^{-1})/2$, $y^o_i = 0$ for $i=m+1,\dots,n$, where u is given by (2.4.24). Hence it is easy to verify that

$$\|y^o\| = u\cdot\beta_m \tag{3.3.20}$$

and that the following system of inequalities is consistent

$$1 + f_j\cdot(\|y^o\| - 2y^o_j) = 1 + u = \|P_o\| \text{ for } j=1,\dots,m$$

$$1 + f_j\cdot\|y^o\| \le 1 + u = \|P_o\| \qquad \text{for } j\ge m+1. \tag{3.3.21}$$

Following Proposition II.4.1 and (3.3.20) we get our claim.

b). Accordingly to (2.4.21) and Proposition II.5.7 corresponding to P_o vector y^o has coordinates $y^o_1 = u\cdot((m-2)\cdot(\beta_m - f_m^{-1}) - f_m^{-1} - 1)/2$, $y^o_j = u\cdot(f_m^{-1} - f_j^{-1})/2$ for $j=2,\dots,m$ and $y^o_j = 0$ for $j\ge m+1$, where u is given by (2.4.24). It is easy to verify that

$$\|y^o\| = f_m^{-1}\cdot u \tag{3.3.22}$$

and that the following system of inequalities is consistent

$$1 + f_j\cdot(\|y^o\| - 2y^o_j) = 1 + u = \|P_o\| \quad \text{for } j=2,\dots,m$$

$$1 + f_j\cdot\|y^o\| \le 1 + u = \|P_o\| \qquad \text{for } j\ge m+1 \tag{3.3.23}$$

By Proposition II.4.1 we get desired result.***

Now we are able to prove the main result of this section.

Theorem III.3.11. Assume $f \in S_{B^*}$, $f=(1,f_2,\dots,f_n)$, $f_s>0$. Then $P_o \in \mathcal{P}(B,D)$ ($D = \ker(f)$) is a unique minimal projection if and only if P_o is a SUBA to 0 in $\mathcal{P}(B,D)$ (we consider the real case).

Proof. Assume P_o is a unique minimal projection and consider the function $\phi: S_D \to \mathbb{R}$ given by

$$\phi(y) = \min(f_{k(g)}\cdot g(y): g \in C) \tag{3.3.24}$$

where $C = \{g \in \text{crit}^* P_o : g(P_o e_i) = \|P_o\| \text{ for some } i \in \{1,\dots,n\}\}$ (3.3.25)

and $k(g) = \min\{i \in \{1,\dots,n\}: g(P_o e_i) = \|P_o\|\}$. (3.3.26)

Assume we can prove that $\phi(y) < 0$ for every $y \in S_D$. Hence, by the compactness of S_D and the continuity of ϕ, the constant $\gamma = \sup(\phi(y): y\in S_D)$ is strictly negative. We will prove that P_o is a SUBA to 0 in $\mathcal{P}(B,D)$ with $r = -\gamma$. To do this, following Theorem 2.5, it is enough to show that for every $P \in \mathcal{P}(B,D)$ there exists $g \in C$ (it is clear that $C \cup -C = \text{crit}^* P_o$

and $C \cap -C = \emptyset$) with $\inf\{g(P-P_o)e_i : e_i \in A_g\} \le -r \cdot \|P-P_o\|$. So fix $P \in \mathcal{P}(B,D)$ and let $P - P_o = f(\cdot)\cdot y$ for some $y \in D$ (we may assume $y \neq 0$). Select $g \in C$ with $f_{k(g)} \cdot g(y/\|y\|) = \phi(y/\|y\|)$. Note that for every $e_i \in A_g$ $g(P-P_o)e_i = f_i \cdot g(y/\|y\|)\cdot\|y\| \ge f_{k(g)} \cdot g(y/\|y\|)\cdot\|y\|$, since $\phi(y/\|y\|) < 0$. Hence $\inf\{g(P-P_o)e_i : i \in A_g\} = f_{k(g)} \cdot g(y) = \phi(y/\|y\|)\cdot\|y\| \le \gamma\cdot\|y\| =$ $= -r\cdot\|P-P_o\|$, which, following Theorem 2.5, gives our assertion.

By the same reasoning as in Theorem 3,1, we can show that the constant r is the best possible.
To end the proof it sufficies to show that $\phi(y) < 0$ for each $y \in S_D$.
By (3.3.24) and (3.3.26) $k(g) \in A$ (see (3.3.19)). Hence following Lemma 3.10 and Theorem II.5.2, we get $f_{k(g)} > 0$. Accordingly to (3.3.24) it is enough to verify that for every $y \in S_D$ $\inf\{g(y) : g \in C\} < 0$. Assume for a contrary that there exists $y \in S_D$ with $g(y) \ge 0$ for every $g \in C$ and consider two cases.

Case I. $a_m > m-2$. If $m = k(f)$ (see Theorem II.4.9) then following Lemma 3.10 the correspopnding to P_o set $A = \{1,\dots,m\}$. Consequently, by Remark 3.9 and (3.3.25) $C = \bigcup_{i=1}^{m} D_i$, where $D_i = \{g \in \text{ext } S_{B^*} : g(P_o e_i) = \|P_o\|\}$.
In view of Proposition II.4.1,

$$D_i = \{(-1,\dots,\underset{i}{1},-1,\dots,\underset{m}{-1},\varepsilon_1,\dots,\varepsilon_{n-m}) : \varepsilon=(\varepsilon_1,\dots,\varepsilon_{n-m}) \in E(n-m)\}.$$

Hence the inequalities $g(y) \ge 0$ for every $g \in C$ form the system (3.3.8). Following Lemma 3.7, we may find $y^1 \in S_D$ with $g(y^1) > 0$ for every $g \in C$. Hence for every $g \in C$ and $e_i \in A_g$

$$f(e_i)\cdot g(y^1) > 0 \tag{3.3.27}$$

since $i \le m$ and $f_m > 0$. Now define $P = P_o + f(\cdot)\cdot y^1$ and note that (3.3.27) yields for every $g \in C$

$$\inf\{g(P-P_o)e_i : e_i \in A_g\} > 0 \tag{3.3.28}$$

Following Theorem 2.5, P_o is not a minimal projection: contradiction.
If $m < k(f)$ then the set $A = \{1,\dots,l\}$, where l is given in Lemma 3.10. Hence $C = \bigcup_{i=1}^{l} D_i$, where for $i=1,\dots,m$ the sets D_i are as above and

$$D_i = \{(-1,\dots,\underset{m}{-1},\varepsilon_1,\dots,\underset{i}{1},\varepsilon_i,\dots,\varepsilon_{n-m-1}) : \varepsilon \in E(n-m-1)\} \text{ for } i=m+1,\dots,l.$$

So to the system (3.3.8) we must add the system

$$y_j \geq \sum_{i=1}^{m} y_i + \sum_{\substack{i=1\\ i\neq j}}^{n-m} |y_{i+m}| \text{ for } j=m+1,\ldots,l.$$

Following Lemma 3.7, there exists $y^1 \in \ker(f)$ with

$$y_j^1 > \sum_{\substack{i=1\\ i\neq j}}^{m} y_i^1 + \sum_{i=1}^{n-m} |y_{i+m}^1| \text{ for } j=1,\ldots,m.$$

Now replace f by $f^1 = (1,f_2,\ldots,f_m,f_{m+1}^1,\ldots,f_n^1)$ where $f_{m+1} > f_{m+1}^1 \geq \ldots \geq f_n^1$. Note that, following Theorem II.4.9, the operator P_o^1 defined by

$$P_o^1 x = x - f^1(x)\cdot y^o \text{ for } x \in B \tag{3.3.29}$$

is a minimal projection from B onto $\ker(f^1)$. If the change of f_{m+1} will be slight, then modifying slightly the n-m last coordinates of vector y^1 we get $y^2 = (y_1^1,\ldots,y_m^1,y_{m+1}^2,\ldots,y_n^2) \in \ker(f^1)$ satisfying (3.3.8). Since $\beta_m < 1/f_{m+1}^1$, applying Theorem 2.5, we get that P_o^1 is not a minimal projection from B onto $\ker(f^1)$; contradiction.

Case II. $a_m < m-2$, $a_{m-1} > m-3$, $f_2 < 1$. If $m < k(f)$ or $m = k(f)$ and $f_{m+1} < f_m$ then by Lemma 3.10, $A = \{2,\ldots,m\}$ and $C = \bigcup_{i=2}^{m} D_i$, where the sets D_i are defined as in the Case I. Following Proposition 4.1,
$D_i = \{(-1,\ldots,-1,\underset{i}{1},-1,\ldots,\underset{m-1}{-1},\varepsilon_1,\ldots,\varepsilon_{n-m+1}): \varepsilon \in E(n-m+1)\}$ for $i=2,\ldots,m-1$ and $D_m = \{(-1,\ldots,-1,\underset{m}{1},\varepsilon_1,\ldots,\varepsilon_{n-m}): \varepsilon \in E(n-m)\}$. Hence the inequalities $g(y) \geq 0$ for every $g \in C$ form system (3.3.13). By Lemma 3.8 there exists $y^1 \in D$ with $g(y^1) > 0$ for every $g \in C$. Reasoniong as in Case I we get contradiction with the minimality of P_o.

If $m = k(f)$ and $f_{m+1} = f_m$, then $C = \bigcup_{i=2}^{l} D_i$, where l is defined in Lemma 3.10 and $D_i = \{(-1,\ldots,\underset{m-1}{-1},\varepsilon_1,\ldots,\varepsilon_{i-1},1,\varepsilon_i,\ldots,\varepsilon_{n-m}): \varepsilon \in E(n-m)\}$ for $i \geq m$ (for $i=2,\ldots,m$ D_i are defined as above). Hence to the system (3.3.13) we must add the following inequalities

$$y_j \geq \sum_{i=1}^{m-1} y_i + \sum_{\substack{i=1\\ i\neq j}}^{n-m} |y_{i+m-1}| \tag{3.3.30}$$

Following Lemma 3.8, there exists $y^1 \in D$ with

$$y_j^1 > \sum_{\substack{i=1\\ i\neq j}}^{m-1} y_i^1 + \sum_{i=1}^{n-m+1} |y_{i+m-1}^1| \quad \text{for } i=2,\ldots,m-1 \text{ and}$$

$$y_m^1 > \sum_{i=1}^{m-1} y_i^1 + \sum_{i=2}^{n-m+1} |y_{i+m-1}^1|.$$

Modifying, as in Case I, f to f^1 where $f^1 = (f_1,\dots f_m, f^1_{m+1},\dots,f^1_n)$, $f^1_{m+1} < f_m$ and y^1 to y^2 belonging to $\ker(f^1)$ we get contradiction as in Case I. The proof of Theorem 3.11 is fully completed.***
Reasoning as in Remark 3.2, we can establish the following

Remark III.3.12. In the complex case Theorem 3.11 does not hold.

§ 4. Criterion for the space $\mathcal{K}(C(T,\mathbb{K}))$

In this section $B = C(T,\mathbb{K})$ the space of all continuous, $\mathbb{K}$-valued functions defined on a compact set T with the supremum norm. For $F \subset T$ by $\mathcal{K}_F(B)$ ($\mathcal{K}(B)$ if $F=T$) we denote the space of all compact operators going from B to B with carriers (see Def.III.1.9) contained in F. For $t \in T$ the symbol $\hat{t}$ stands for the evaluation functional. We start with the following

Lemma III.4.1. Assume that $V \in \mathcal{K}(B)\setminus\{0\}$ and let card car$(V) < \infty$, i.e. $V \in \mathcal{D}(B)$. For $\hat{t} \in \text{crit}^*V$ (see 3.2.1) put

$$A_t = \{x \in S_B : (Vx)t = \|V\|\}. \qquad (3.4.1)$$

Then for every $\hat{t} \in \text{crit}(V)$ and every $\{x_n\} \subset S_B$ with $(Vx_n)t \to \|V\|$, there exists $\{z_n\} \subset A_t$ with $\|z_n - x_n\| \to 0$ as $n \to \infty$.

Proof. Since $V \in \mathcal{D}(B)$, $V = \sum_{i=1}^{k} \hat{t}_i(\cdot)\cdot y_i$, where $y_i \in B$, $t_i \in T$ for $i=1,\dots,k$.

By the Tietze-Urysohn Theorem $\|V\| = \|\sum_{i=1}^{k} |y_i|\,\|$. Fix $\hat{t} \in \text{crit}^*V$, $\{x_n\} \subset S_B$ with $(Vx_n)t \to \|V\|$ and let $A = \{i \in \{1,\dots,k\} : y_i(t) \neq 0\}$.

At first we will show that $x_n(t_i) \to \overline{y_i(t)}/|y_i(t)| = \text{sgn}(y_i(t))$ for $i \in A$.

Since $\sum_{i=1}^{k} |y_i(t)| = \sum_{i\in A} \text{sgn}(y_i(t))\cdot y_i(t)$, $|x_n(t_i)| \to 1$ for each $i \in A$.

Assume that for some $i_o \in A$ there exists a subsequence (x_{n_k}) with

$$|\text{sgn}(y_{i_o}(t)) - x_{n_k}(t_{i_o})/|x_{n_k}(t_{i_o})|| \geq d > 0 \text{ for } k \geq k_o.$$

By the uniform convexity of $\mathbb{C}$ over $\mathbb{R}$,

$$|\tfrac{1}{2}\cdot(\text{sgn}(y_{i_o}(t)) + x_{n_k}(t_{i_o})/|x_{n_k}(t_{i_o})|)| \leq 1-\delta \text{ for some } \delta > 0.$$

Compute

$$\left|\tfrac{1}{2}\cdot\left(\sum_{i\in A}|y_i(t)| + \sum_{i\in A}((x_{n_k}(t_i)/|x_{n_k}(t_i)|)\cdot y_i(t))\right)\right| \le$$

$$\le \sum_{i\in A\setminus\{i_o\}}|y_i(t)| + \left|\tfrac{1}{2}\cdot(\mathrm{sgn}(y_{i_o}(t)) + x_{n_k}(t_{i_o})/|x_{n_k}(t_{i_o})|)\right|\cdot|y_{i_o}(t)| \le$$

$$\le \sum_{i\in A\setminus\{i_o\}}|y_i(t)| + (1-\delta)\cdot|y_{i_o}(t)| < \|V\|.$$

But, passing to the subsequence if necessary, $\sum_{i\in A}(x_{n_k}(t_i)/|x_{n_k}(t_i)|)\cdot y_i(t)$ tends to $\|V\|$ as $k \to \infty$; contradiction.

Now we construct the sequence (z_n). For each $n \in \mathbb{N}$ let us set

$$\varepsilon_n = \max\{|x_n(t_i)-\mathrm{sgn}(y_i(t))| : i \in A\}.$$

Fix $n \in \mathbb{N}$ and for every $i \in A$ select an open neighbourhood U_i of t_i such that $U_i \cap \overline{U}_j = \emptyset$ for $i\neq j$ and $|x_n(s) - x_n(t_i)| \le \varepsilon_n$ for $s \in U_i, i \in A$. Fix $i \in A$. An easy calculation shows that for every $s \in \overline{U}_i$

$$\mathrm{re}(x_n(s)) \in [\mathrm{re}(\mathrm{sgn}(y_i(t)))-2\cdot\varepsilon_n, \mathrm{re}(\mathrm{sgn}(y_i(t)))+2\cdot\varepsilon_n] \cap [-1,1]=[B,C]$$

and

$$\mathrm{im}(x_n(s)) \in [\mathrm{im}(\mathrm{sgn}(y_i(t)))-2\cdot\varepsilon_n, \mathrm{im}(\mathrm{sgn}(y_i(t)))+2\cdot\varepsilon_n] \cap [-1,1]=[D,E]$$

Let us set $S_i = \sigma(U_i) \cup \{t_i\}$ and define for $s \in S_i$

$$f_i(s) = \begin{cases} \mathrm{re}(x_n(s)) & ; s \in \sigma(U_i) \\ \mathrm{re}(\mathrm{sgn}(y_i(s))) & ; s=t_i \end{cases} \quad \text{and}$$

$$g_i(s) = \begin{cases} \mathrm{im}(x_n(s)) & ; s \in \sigma(U_i) \\ \mathrm{im}(\mathrm{sgn}(y_i(s))) & ; s=t_i \end{cases}.$$

Following the Tietze-Urysohn Theorem, we can extend in a continuous way the functions f_i and g_i on the whole set $\overline{U}_i$ such that $f_i(s) \in [B,C]$ and and $g_i(s) \in [D,E]$ for every $s \in \overline{U}_i$. It is easy to show that

$$|(f_i + i\cdot g_i)(s) - \mathrm{sgn}(y_i(t))| \le 2\cdot\sqrt{2}\cdot\varepsilon_n.$$

Let $\pi_i : B_d(\mathrm{sgn}(y_i(t)), \sqrt{2}\cdot 2\cdot\varepsilon_n) \to B_d(\mathrm{sgn}(y_i(t)), \sqrt{2}\cdot 2\cdot\varepsilon_n) \cap B_d(0,1)$ $(B_d(x,r) = \{y \in \mathbb{C}: |x-y|\le r\})$ be a continuous function with $\pi_i|_{B_d(\mathrm{sgn}(y_i(t)),r)\,\cap\, B_d(0,1)} = \mathrm{id}$ $(r = \sqrt{2}\cdot 2\cdot\varepsilon_n)$.

Put $z_i^n = \pi_i \circ (f_i + i\cdot g_i)$. We note that z_i^n is continuous, $z_i^n(t_i) = \mathrm{sgn}\, y_i(t)$ and $\sup\{|z_i^n(s)| : s \in \overline{U}_i\} = 1$. Now define a function $z_n : T \to \mathbb{C}$ by:

$$z_n(s) = \begin{cases} x_n(s) : s \in T \setminus \bigcup_{i\in A} \overline{U}_i \\ z_i^n(s) : s \in \overline{U}_i \end{cases}.$$

Since for every $i \in A$ and $s \in \sigma(U_i)$ $z_n(s) = x_n(s)$, z_n is continuous. Moreover $\|z_n\| = 1$ and $z_n(t_i) = \operatorname{sgn} y_i(t)$ for $i \in A$, which means that $z_n \in A_t$.

To finish the proof, it is sufficient to show that $\|z_n - x_n\| \to 0$. Fix $s \in T$. If $s \in T \setminus \bigcup_{i\in A} U_i$, then $|(x_n - z_n)(s)| = 0$. If $s \in \overline{U}_i$ for some $i \in A$, then $|x_n(s) - z_n(s)| \le |x_n(s) - x_n(t_i)| + |x_n(t_i) - \operatorname{sgn}(y_i(t))| +$

$+ |\operatorname{sgn}(y_i(t)) - z_n(s)| \le (2+\sqrt{2}\cdot 2)\cdot\varepsilon_n$.

But it gives that $\|z_n - x_n\| \to 0$, since $\varepsilon_n \to 0$. The proof is completed.***

Now we will prove the main result of this section.

Theorem III.4.2. Let $\mathcal{V} \subset \mathcal{K}_F(B)$ be a convex set. Take $K \in \mathcal{K}_F(B)$, $V \in \mathcal{V}$ and assume $K-V \in \mathcal{D}(B)$. Then we have:

(a) $V \in P_{\mathcal{V}}(K)$ if and only if for every $U \in \mathcal{V}$ there exists $\hat{t} \in \operatorname{crit}^* K-V$ such that $\inf\{\operatorname{re}(((U-V)x)t) : x \in A_t\} \le 0$, where A_t is defined by (3.4.1);

(b) V is a SUBA to K in $\mathcal{V}$ with a constant $r>0$ if and only if for for every $U \in \mathcal{V}$ there exists $\hat{t} \in \operatorname{crit}^* K-V$ such that $\inf\{\operatorname{re}(((U-V)x)t) : x \in A_t\} \le -r\cdot\|U-V\|$.

Proof. a) Assume that $V \notin P_{\mathcal{V}}(K)$. Then there exists $U \in \mathcal{V}$ with $\|K-U\| < \|K-V\|$. Take $\hat{t} \in \operatorname{crit}(K-V)$ and $x \in A_t$. We note that $\operatorname{re}(((U-V)x)t) =$

$= \operatorname{re}(((K-V)x)t) - \operatorname{re}(((K-U)x)t) \ge \|K-V\| - \|K-U\| > 0$ and consequently $\inf\{\operatorname{re}((U-V)x)t) ; x \in A_t\} > 0$.

To prove the converse suppose that for some $U \in \mathcal{V}$ and every $\hat{t} \in \operatorname{crit}(K-V)$ $\inf\{\operatorname{re}(((U-V)x)t) : x \in A_t\} > 0$. Following Theorem (1.1) it is sufficient to show that $\operatorname{re}(f(U-V)) > 0$ for every $f \in E(K-V)$ (see (3.1.3)). So fix $f \in E(K-V)$. By Theorem (III.1.8) and Corollary (1.10) $f = x^{**} \otimes \hat{t}$ for some $t \in T$ and $x^{**} \in \operatorname{ext} S(B^{**})$. Applying Goldstine's Theorem we may select a net $\{x_\beta\} \subset S_B$ tending weak* in B^{**} to x^{**}. Following (3.1.23), we note that

$$\|K-V\| \ge \operatorname{re}(((K-V)x_\beta)t) \to \operatorname{re}(((K-V)^* x^{**})t) = \operatorname{re}(f(K-V)) = \|K-V\|$$

and consequently $\hat{t} \in \operatorname{crit}(K-V)$.

Now let us set $f_\beta = x_\beta \otimes \hat{t}$ and observe that for every $W \in \mathcal{K}(B)$

$$f_\beta(W) = \hat{t}(W(x_\beta)) \to \hat{t}(W^*(x^{**})) = (x^{**} \otimes \hat{t})(W^*) = f(W).$$

Hence we may select a sequence $\{f_n\} \subset \{f_\beta\}$ $(f_n = x_n \otimes \hat{t})$ such that $f_n(K-V) \to f(K-V) = \|K-V\|$ and $f_n(U-V) = ((U-V)x_n)t) \to f(U-V)$. Following Lemma (4.1), there exists a sequence $\{z_n\} \subset A_t$ with $\|z_n - x_n\| \to 0$. It is clear that $((U-V)z_n - (U-V)x_n)t \to 0$ which yields $((U-V)z_n)t \to f(U-V)$. Since for $n=1,2,\ldots$ $z_n \in A_t$ and $t \in crit(K-V), re(f(U-V)) > 0$ which accordingly to Theorem (1.1) completes the proof of part (a).
Applying Theorem (1.3), part (b) can be shown in the same way.***

Corollary III.4.3. Assume $\mathcal{V}, \mathcal{K}_F(B), K, V$ are the same as in Theorem (4.2). Assume furthermore that card $F < +\infty$. Then we have:

(a) $V \in P_{\mathcal{V}}(K)$ if and only if for every $U \in \mathcal{V}$ there exists $\hat{t} \in crit^* K-V$ with $\inf\{re(((U-V)x)t) : x \in A_t\} \le 0$.

(b) V is a SUBA to K in $\mathcal{V}$ with a constant $r>0$ if and only if for every $U \in \mathcal{V}$ there exists $\hat{t} \in crit^* K-V$ with $\inf\{re(((U-V)x)t) : x \in A_t\} \le -r \cdot \|U-V\|$.

Theorem (4.2) yields immediately the following result:

Theorem III.4.4. Let $D \subset B$ be its n-dimensional subspace and let $\mathcal{V} = \mathcal{P}(B,D)$. Assume $P_o \in \mathcal{P}(B,D) \cap \mathcal{D}(B,D)$. Then P_o is minimal in $\mathcal{P}(B,D)$ (resp. P_o is a SUBA to 0 in $\mathcal{P}(B,D)$ with a constant $r>0$) if and only if for every $P \in \mathcal{P}(B,D)$ there exists $\hat{t} \in crit^* P_o$ such that $\inf\{re(((P_o - P)x)t) : x \in A_t\} \le 0$ (resp. $\le -r \cdot \|P-P_o\|$).

Proof. Take $K = 0, V = P_o$ and note that $crit^* P_o = crit^* -P_o$. By Theorem (4.2), we derive the desired result.***

We note that Theorem (4.4) extends the result of Cheney (see [38]) proved for $P_o \in I(B,D)$ (see (3.1.25)) in the real case.

§ 5. Applications to the set $\mathcal{P}_D(C(T,\mathbb{K}),D,F)$

In this section we apply Theorem III.4.2 to obtain some criteria for the case of projections with finite carriers. At first we introduce some notions. In this section $B = C(T,\mathbb{K})$ ($\mathbb{K}=\mathbb{C}$ or $\mathbb{K}=\mathbb{R}$). Let $D \subset B$, dim $D=n$ and let $F = \{t_1, \dots, t_m\}, t_i \neq t_j$ for $i \neq j, m \geq n+1$. Assume furthermore that F is total over D i.e. if $y \in D$, $y(t_j) = 0$ for $j=1, \dots, m$ then $y = 0$. Since $\dim D = n$, we may numerate the points from F in such a way that $(\hat{t}_1|_D, \dots, \hat{t}_n|_D)$ form a basis of D^*. For $i=n+1, \dots, m$ put $B_i = \{1, \dots, n, i\}$ and select for $j \in B_i$ the numbers τ_i^j such that

$$\sum_{j \in B_i} |\tau_i^j| > 0 \text{ and } \sum_{j \in B_i} (\tau_i^j \cdot \hat{t}_j)|_Y = 0. \tag{3.5.1}$$

Let us assume $P \in \mathcal{P}_D(B, D, F)$ (see section 1), $P = \sum_{j=1}^{m} \hat{t}_j(\cdot) \cdot u_j$, where $u_j \in D$ for $j=1, \dots, m$. For $i=n+1, \dots, m$ define the functions $v_i^P : T \to \mathbb{C}$ by

$$v_i^P(s) = \sum_{j \in B_i} \tau_i^j \cdot \operatorname{sgn} u_j(s) \tag{3.5.2}$$

and the functionals ϕ_i by

$$\phi_i = \sum_{j \in B_i} \tau_i^j \cdot \hat{t}_j. \tag{3.5.3}$$

Then we can prove the following

Theorem III.5.1.

(a) P is not a minimal projection in $\mathcal{P}_D(B, D, F)$ if and only if for every $i \in \{n+1, \dots, m\}$ there exist $y_i \in D$ such that for every $\hat{s} \in \operatorname{crit}^* P$

$$\operatorname{re}\Big(\sum_{i=n+1}^{m} v_i^P(s) \cdot y_i(s) - \sum_{j \in B_s^P} \Big|\sum_{i=n+1}^{m} \tau_i^j \cdot y_i(s)\Big| - \sum_{j \in C_s^P} |y_j(s) \cdot \tau_j^j|\Big) > 0 \tag{3.5.4}$$

where $B_s^P = \{j \in \{1, \dots, n\} : u_j(s) = 0\}$, $C_s^P = \{j \in \{n+1, \dots, m\} : u_j(s) = 0\}$,

$\sum_{j \in B_s^P} = 0$ (resp. $\sum_{j \in C_s^P} = 0$) if $B_s^P = \emptyset$ (resp. $C_s^P = \emptyset$).

(b) P is not a SUBA to 0 in $\mathcal{P}_D(B, D, F)$ with a constant $r > 0$ if and only if for every $i=n+1, \dots, m$ there exists $y_i \in D$ such that for every $\hat{s} \in \operatorname{crit}^* P$

$$\operatorname{re}\Big(\sum_{i=n+1}^{m} v_i^P(s) \cdot y_i(s) - \sum_{j \in B_s^P} \Big|\sum_{i=n+1}^{m} \tau_i^j \cdot y_i(s)\Big| - \sum_{j \in C_s^P} |\tau_j^j \cdot y_j(s)|\Big) >$$

$$> -r \cdot \|L\|, \tag{3.5.5}$$

where $L = \sum_{i=n+1}^{m} \phi_i(\cdot) \cdot y_i$.

Proof.(a) Assume that condition (3.5.4) is fulfilled and let

$L = \sum_{i=n+1}^{m} \phi_i(\cdot)\cdot y_i$. To prove that P is not a minimal projection, in view of Theorem III.4.2, it is sufficient to show that for $\hat{e}$ach $s \in crit^* P$

$$\inf\{re((Lx)s) : x \in A_s\} > 0.$$

Let us denote for $i=n+1,\dots,m$ $D_i=\{j \in B_i : u_j(s)\neq 0\}$ and $E_i = B_i \setminus D_i$. Fix $s \in crit^* P$, $x \in A_s$ and compute

$$(Lx)s = \sum_{i=n+1}^{m} \phi_i(x)\cdot y_i = \sum_{i=n+1}^{m} \Big(\sum_{j=1}^{n} \tau_i^j \cdot x(t_j) + \tau_i^i \cdot x(t_i) \Big) =$$

$$= \sum_{i=n+1}^{m} \Big(\sum_{j\in D_i} \tau_i^j \cdot \mathrm{sgn}(u_j(s)) - \sum_{j\in E_i} \tau_i^j \cdot (-x(t_j)) \Big) =$$

$$= \sum_{i=n+1}^{m} v_i^P(s)\cdot y_i(s) - \sum_{i=n+1}^{m} \Big(\sum_{j\in E_i} \tau_i^j \cdot (-x(t_j)\cdot y_i(s) \Big) =$$

$$= \sum_{i=n+1}^{m} v_i^P(s)\cdot y_i(s) - \sum_{j\in B_s^P} \Big(\sum_{i=n+1}^{m} \tau_i^j \cdot y_i(s)) \cdot (-x(t_j)) - \sum_{j\in C_s^P} \tau_j^j \cdot y_j(s) \cdot (-x(t_j))$$

Consequently, since $\|x\| \le 1$, we obtain

$$re((Lx)s) \ge re\Big(\sum_{i=n+1}^{m} v_i^P(s)\cdot y_i(s) - \sum_{j\in B_s^P} \Big| \sum_{i=n+1}^{m} \tau_i^j \cdot y_i(s) \Big| -$$

$$- \sum_{j\in C_s^P} |\tau_j^j \cdot y_j(s)| \Big) > 0.$$

By Theorem III.4.2, P is not a minimal projection in $\mathcal{P}_D(B,D,F)$. To prove the converse, assume P is not minimal in $\mathcal{P}_D(B,D,F)$ and choose $P_o \in \mathcal{P}_D(B,D,F)$ with $\|P_o\| < \|P\|$. By ([42]), we may assume

$$P_o = P + \sum_{i=n+1}^{m} \phi_i(\cdot)\cdot y_i \text{ for some } y_{n+1},\dots,y_m \in D.$$

We show that the functions $y_{n+1},\dots,y_m$ satisfy (3.5.4). Fix $\hat{s} \in crit^* P$. By the Tietze-Urysohn Theorem we may define a function $x \in S_D$ with the properties

$$x(t_j) = \begin{cases} \mathrm{sgn}(u_j(s)) & ; u_j(s)\neq 0 \\ -\mathrm{sgn}\Big(\sum_{i=n+1}^{m} \tau_i^j \cdot y_i(s)\Big) & ; u_j(s)=0 \end{cases} \quad \text{for } j=1,\dots,n$$

and

$$x(t_j) = \begin{cases} \mathrm{sgn}(u_j(s)) & ; u_j(s)\neq 0 \\ -\mathrm{sgn}(\tau_j^j \cdot y_j(s)) & ; u_j(s)=0 \end{cases} \quad \text{for } j=n+1,\dots,m.$$

Observe that

$$(Px)s = \sum_{j=1}^{m} x(t_j)\cdot u_j(s) = \sum_{j\in B_s^P \cup C_s^P} x(t_j)\cdot u_j(s) = \sum_{j=1}^{m} |u_j(s)| = \|P\|.$$

Calculating as in the previous part of the proof we obtain

$$((P_o - P)x)s = \sum_{i=n+1}^{m} v_i^P(s)\cdot y_i(s) - \sum_{j\in B_s^P} \Big|\sum_{i=n+1}^{m} \tau_i^j\cdot y_i(s)\Big| - \sum_{j\in C_s^P} |\tau_j^j\cdot y_j(s)|.$$

Since, following Theorem III.4.2, $re(((P_o - P)x)s)) > 0$, the proof of part (a) is fully completed.

The proof of part (b) goes on the same line, so we omit it. ***

Observe that in the real case if $m=n+1$ condition (3.5.2) reduces to the following

$$|y_{n+1}(s)|\cdot(v_{n+1}^P(s)\cdot \operatorname{sgn} y_{n+1}(s) - \sum_{j\in B_s^P \cup C_s^P} |\tau_{n+1}^j|) > 0 \tag{3.5.6}$$

which after dividing by $|y_{n+1}(s)|$ yields the result of Cheney (see[42], Th. 5).

Now we specialize our results to the case $T = [a,b]$, D being an n-dimensional Haar subspace of $B = C(T,\mathbb{R})$ and $F = (t_1,\dots,t_{n+1})$. Define

$$E = \sup(\|y\|: y \in D \ |y(t)| \le 1 \text{ for } t \in F) \tag{3.5.7}$$

and for $P \in \mathcal{P}_D(B,D,F)$, $s \in T$ put

$$u^P(s) = \sum_{j\in B_s^P \cup C_s^P} |\tau_{n+1}^j|. \tag{3.5.8}$$

At first we after [42] we present without proof some preliminary lemmas.

Lemma III.5.2. If $s \in T$, $y \in S_B$, $(\hat{s}\circ P)y = \|\hat{s}\circ P\|$, and $|v^P(s)| > u^P(s)$, then $\Phi(y)\cdot v^P(s) > 0$. (we write Φ instead of Φ_{n+1} and v^P instead of v_{n+1}^P).

Lemma III.5.3. If $\|P\| > E$ for $P \in \mathcal{P}_D(B,D,F)$ then $|v^P(s)| > u^P(s)$ for every $\hat{s} \in \operatorname{crit}^* P$.

Lemma III.5.4. $\|P\| \ge E$ for every $P \in \mathcal{P}_D(B,D,F)$.

Lemma III.5.5. Assume $F \subset [a,b]$, F is closed and let $D \subset B$ be its n-dimensional Haar subspace. Let $\gamma: F \to \mathbb{R}$ be a function which has no roots and such that $\operatorname{sgn} \gamma$ is continuous. If no $y \in D$ has the property $\gamma(y)|_F > 0$ then there exist $n+1$ points $t_1,\dots,t_{n+1}$ in F such that $t_1<t_2<\dots<t_{n+1}$ and $\gamma(t_{i-1})\cdot\gamma(t_i)<0$ for $i=2,\dots,n+1$.

Now we are able to prove the following

Theorem III.5.6. In order that P be minimal in the set $\mathcal{P}(B,D,F)$ it is necessary and sufficient that either

a) $\|P\| = E$; or

b) there exist $s_1 < s_2 < \ldots < s_{n+1}$ with $\hat{s}_i \in \text{crit}^* P$ for $i=1,\ldots,n+1$ such that $\text{sgn}\ v^P(s_i) = -\text{sgn}\ v^P(s_{i-1})$ $(2 \le i \le n+1)$.

Proof. First suppose that a) holds. By Lemma 5.4, P is a minimal projection in $\mathcal{P}_D(B,D,F)$.

Next suppose that b) holds. Then no element $y \in D$ can have the property $v^P(s)\cdot \text{sgn}\ y(s) > u^P(s)$ for $\hat{s} \in \text{crit}^* P$, for this inequality would require y to have at least n roots. Hence, by (3.5.6) and Theorem 5.1, P is minimal.

Finally suppose that P is minimal in $\mathcal{P}_D(B,D,F)$ and a) is not true. Then $\|P\| > E$. Following Theorem 5.1 and (3.5.6), no $y \in D$ satisfies $v^P(s)\cdot \text{sgn}\ y(s) > u^P(s)$ for any $\hat{s} \in \text{crit}^* P = S$. By Lemma 5.3, $|v^P| > u^P$ on S. Hence no y in D can satisfy the inequality $y\cdot v^P > 0$ on S. Since $u^P \ge 0$, v^P does not vanish on S. Now we verify that sgn v^P is continuous on S. If sgn v^P is discontinuous on S then consider two sets

$S_1 = \{t \in S\colon\ v^P(s) > 0\}$,

$S_2 = \{t \in S\colon\ v^P(s) < 0\}$.

One of these sets must contain an accumulation point of the other. But this is not possible, for as we show, S_1 and S_2 are closed. Consider, for example S_1. For each $\varepsilon \in E(n+1)$ and $t \in S_1$ select $z_\varepsilon \in S_B$ with

$$(\hat{t}\circ P)z_\varepsilon = \|P\| \text{ and } \sum_{i=1}^{n+1} z_\varepsilon(t_i)\cdot\tau_i^{n+1} = v^P(t) > 0. \qquad (3.5.9)$$

Put $Z = \{z_\varepsilon : \varepsilon \in E(n+1)\}$. Note that for each $z \in Z$ the set of all $t \in [a,b$ satisfying (3.3.9) is closed. Since S_1 is the union of such sets and the set Z is finite, S_1 is closed too; contradiction with discontinuity of γ. Following Lemma 5.5, the proof is completed***

Theorem III.5.7. If P_o is a minmal projection in $\mathcal{P}_D(B,D,F))$ and if $\|P_o\| >$ $> E$ then P_o is a SUBA to 0 in $\mathcal{P}_D(B,D,F)$.

Proof. Take any $P \in \mathcal{P}(B,D,F)$. Following [42] P admits a representation $P = P_o + \Phi(\cdot)\cdot y$, where $y \in D$ (we write Φ instead of Φ_{n+1}). It is evident that $\|P - P_o\| = \|y\|$. By Theorem 5.6, there exist $\hat{s}_1,\ldots,\hat{s}_{n+1}$ from the set $\text{crit}^* P_o$ such that $v^P(s_i)\cdot v^P(s_{i-1}) < 0$ for $i=2,\ldots,n+1$. For each i, select $z_i \in Z$ (Z is as in the previous theorem) with $(\hat{s}_i \circ P_o)z_i = \|P\|$. By Lemma 5.3, $|v^P(s_i)| > u^P(s_i)$. By Lemma 5.6, $\Phi(z_i) \cdot v^P(s_i) > 0$. Hence $\|P\| \geq \|P_o\| + \Phi(z_i) \cdot y(s_i)$ for $i=1, \ldots, n+1$. Define

$$r = \inf\{\max\{\Phi(z_i) \cdot w(s_i) : i=1, \ldots, n+1\} : w \in S_D\} \tag{3.5.10}$$

Since $v^P(s_i)$ alternates in sign, so does $\Phi(z_i)$. Since $w(s_i)$ cannot, $r>0$. Thus $\|P\| \geq \|P_o\| + \max\{\Phi(z_i) \cdot y(s_i) ; i=1, \ldots, n+1\} \geq \|P_o\| + r \cdot \|y\|$. The proof is completed.***

Theorem III.5.8. If $\|P_o z\| < E$ and $\Phi(z) \neq 0$, then P_o is a minimal projection in $\mathcal{P}_D(B,D,F)$ but it is not unique.

Proof. Since the set Z (defined in the proof of Theorem 5.6) is finite, the number $\varepsilon = E - \max\{\|P_o z\| : z \in Z, \Phi(z) \neq 0\}$ is positive. Since $\|P_o\| = \max\{\|P_o z\| : z \in Z\}$, we conclude

$$E \leq \|P_o\| = \max\{\|P_o z\| : z \in Z, \Phi(z)=0\} \leq \max\{\|P_o z\| : z \in Z, (P_o z)t_i = z(t_i)\}$$

$$= \max\{\|y\| ; y \in D, |y(t_i)|=1\} \leq E.$$

Now if $\|y\| \leq \varepsilon$ and $\Phi(z) \neq 0$, then

$$\|P_o z + \Phi(z) \cdot y\| \leq \|P_o z\| + \|y\| \leq (E - \varepsilon) + \varepsilon = E.$$

If $\Phi(z)=0$, then $\|P_o z + \Phi(z) \cdot y\| = \|P_o z\| \leq E$. Consequently for every $y \in W_d(0,\varepsilon)$, $\|P_o + \Phi(\cdot) \cdot y\| = E$. By Lemma 5.8, the proof is completed.***

§ 6. The case of sequence spaces

Assume $D \subset c_o$ (see section 1) is a n-dimensional subspace and let $y_1, \ldots, y_n$ be a basis of D. For $K \in \mathcal{K}(c_o, D)$, $K = \sum_{i=1}^{n} f_i(\cdot) \cdot y_i$ ($f_i \in l_1$) put

$$K_K(s,t) = \sum_{i=1}^{n} f_i(s) \cdot y_i(t) \quad \text{for } s,t \in T. \tag{3.6.1}$$

As in the previous section for $t \in T$ the symbol $\hat{t}$ stands for the evaluation functional. By ([45], Lemma 1)

$\hat{t} \in \operatorname{crit}^* K$ if and only if t is a critical point of the function $\Lambda_K : T \to \mathbb{R}_+$ defined by $\Lambda_K(s) = \| \sum_{i=1}^{n} y_i(s) \cdot f_i \|_1 =$

$= \sum_{u \in T} |K_K(u,s)|$ i.e. $\Lambda_K(t) = \sup\{\Lambda_K(s) : s \in T\}$ (the symbol

$\|\cdot\|_1$ denotes the norm in the space l_1). (3.6.2)

Using these notations we may prove the following

Theorem III.6.1. Let $\mathcal{V} \subset \mathcal{K}(c_o, D)$ be a convex set and let $K \in \mathcal{K}(c_o, D)$, $V \in \mathcal{V}$. Then we have:

(a) $V \in P_{\mathcal{V}}(K)$ if and only if for every $U \in \mathcal{V}$ there exists $\hat{t} \in \mathrm{crit}^*(K-V)$ with

$$\mathrm{re}\Big(\sum_{s \in T} K_{U-V}(s,t) \cdot \mathrm{sgn}\, K_{K-V}(s,t) \quad - \sum_{s \in A_t} |K_{U-V}(s,t)|\Big) \le 0. \qquad (3.6.3)$$

(b) $V \in \mathcal{V}$ is a SUBA to K in $\mathcal{V}$ with a constant $r>0$ if and only if for every $U \in \mathcal{V}$ there exists $\hat{t} \in \mathrm{crit}^*(K-V)$ such that

$$\mathrm{re}\Big(\sum_{s \in T} K_{U-V}(s,t) \cdot \mathrm{sgn}(K_{K-V}(s,t)) - \sum_{s \in A_t} |K_{U-V}(s,t)|\Big) \le -r \cdot \|U-V\|, \quad (3.6.4)$$

where $A_t = \{s \in T\colon K_{K-V}(s,t)=0\}$.

Proof. Assume there exists $U \in \mathcal{V}$ such that for every $\hat{t} \in \mathrm{crit}^*(K-V)$ (3.6.2) does not hold. In view of Theorem (III.1.1), it is sufficient to show that $\mathrm{re}(\phi(U-V)) > 0$ for every $\phi \in E(K-V)$ (see (3.1.3)) .Since $\mathcal{K}(c_o, D) \subset \mathcal{K}(c_o)$, by Corollary (1.10) $\phi = \psi \otimes \gamma$ for some $\psi \in \mathrm{ext}\, S_{c_o^{**}}$ and $\gamma \in \mathrm{ext}\, S_{c_o^*}$. Applying (3.1.27) and (3.1.28), we may assume that $\psi \in l_\infty$ $|\psi(s)| = 1$ for every $s \in T$ and $\gamma = \hat{t}$ for some $t \in T$. Let

$$K-V = \sum_{i=1}^{n} f_i(\cdot) \cdot y_i \text{ and } U-V = \sum_{i=1}^{n} g_i(\cdot) \cdot y_i \text{ for some } f_i, g_i \in l_1.$$

Following Remark (1.7) and (3.6.2) we note that

$$\|K-V\| = \phi(K-V) = \hat{t}(K-V)^*\psi = \sum_{i=1}^{n} \psi(f_i) \cdot y_i(t) =$$

$$= \sum_{i=1}^{n} \Big(\sum_{s \in T} f_i(s) \cdot \psi(s)\Big) \cdot y_i(t) = \sum_{s \in T} \psi(s) \cdot \Big(\sum_{i=1}^{n} f_i(s) \cdot y_i(t)\Big) \le$$

$$\le \sum_{s \in T} |K_{K-V}(s,t)| = \|K-V\|.$$

It means that $\psi(s) = \mathrm{sgn}(K_{K-V}(s,t))$ if $s \in T \setminus A_t$. Compute

$$\mathrm{re}(\phi(U-V)) = \mathrm{re}\Big(\sum_{i=1}^{n} \psi(g_i) \cdot y_i(t)\Big) = \mathrm{re}\Big(\sum_{i=1}^{n}\Big(\sum_{s \in T} \psi(s) \cdot g_i(s)\Big) \cdot y_i(t)\Big) =$$

$$= \mathrm{re}\Big(\sum_{s \in T} \psi(s) \cdot \Big(\sum_{i=1}^{n} g_i(s) \cdot y_i(t)\Big)\Big) = \mathrm{re}\Big(\sum_{s \in T} \psi(s) \cdot K_{U-V}(s,t)\Big) =$$

$$= \mathrm{re}\Big(\sum_{s \in T} K_{U-V}(s,t) \cdot \mathrm{sgn}(K_{K-V}(s,t)) - \sum_{s \in A_t} (-\psi(s)) \cdot K_{U-V}(s,t)\Big).$$

Since $\Big|\mathrm{re}\Big(\sum_{s \in A_t} (-\psi(s)) \cdot K_{U-V}(s,t)\Big)\Big| \le \sum_{s \in A_t} |K_{U-V}(s,t)|$,

$$\mathrm{re}(\phi(U-V)) \ge \mathrm{re}\Big(\sum_{s\in T} K_{U-V}(s,t)\cdot \mathrm{sgn}(K_{K-V}(s,t))\Big) - \sum_{s\in A_t} |K_{U-V}(s,t)| > 0.$$

Following Theorem (1.1), $V \notin P_{\mathcal{V}}(K)$.

To prove the converse, suppose $V \notin P_{\mathcal{V}}(K)$ and choose $U \in \mathcal{V}$ with $\|U-K\| < \|V-K\|$. Let $\hat{t} \in \mathrm{crit}^*(K-U)$ be fixed. Define a function $\psi \in l_\infty$ by

$$\psi(s) = \begin{cases} \mathrm{sgn}(K_{K-V}(s,t)) & ; K_{K-V}(s,t) \neq 0 \\ -\mathrm{sgn}(K_{U-V}(s,t)) & ; K_{K-V}(s,t) = 0, K_{U-V}(s,t) \neq 0. \\ 1 & ; \text{in the opposite case} \end{cases}$$

Let us set $\phi = \psi \otimes \hat{t}$. Following ([118]), $\phi \in \mathrm{ext}\, S_{\mathcal{K}^*(c_0)}$. Observe that

$$\phi(K-V) = \sum_{i=1}^{n} \psi(f_i)\cdot y_i(t) = \sum_{i=1}^{n}\Big(\sum_{s\in T}\psi(s)\cdot f_i(s)\Big)y_i(t) =$$

$$= \sum_{s\in T}\psi(s)\cdot\Big(\sum_{i=1}^{n} f_i(s)\cdot y_i(t)\Big) = \sum_{s\in T}|K_{K-V}(s,t)| = \|K-V\|.$$

Hence $\phi \in E(K-V)$ and, by Theorem (1.1), $\mathrm{re}(\phi(U-V)) > 0$. But

$$\mathrm{re}(\phi(U-V)) = \mathrm{re}\Big(\sum_{s\in T}\psi(s)\cdot K_{U-V}(s,t)\Big) = \mathrm{re}\Big(\sum_{s\in T} K_{U-V}(s,t)\cdot \mathrm{sgn}(K_{K-V}(s,t))\Big) -$$

$$- \sum_{s\in A_t} |K_{U-V}(s,t)|,$$

which gives the desired result.

Following Theorem (1.3), part (b) can be proved in the same way. ***

Remark III.6.2. In the real case for $K = 0$ and $\mathcal{V} = \mathcal{P}(c_0,D)$ Theorem 6.1 a) was proved by a different method in ([45],Th.1).

Now we present a similar result for the space $\mathcal{K}(l_1,D)$. To do this, for $K \in \mathcal{K}(l_1,D)$, $K = \sum_{i=1}^{n} f_i(\cdot)\cdot y_i$, where $f_i \in l_\infty$ for $i=1,\dots,n$ and $y_1,\dots,y_n$ is a fixed basis of D, put

$$K_K(\psi,t) = \sum_{i=1}^{n}\psi(f_i)\cdot y_i(t), \psi \in l_1^{**},\ t \in T. \tag{3.6.5}$$

Following the Banach-Alaoghlu and the Krein-Milman Theorems, and by the definition of the space $\mathcal{L}_e(l_1^{**},D)$ (see Remark 1.7), we note that the set

$$C_K = \{\psi \in \mathrm{ext}(S(l_1^{**})) : K^*(\psi) = \|K\|\} \tag{3.6.6}$$

is nonvoid. Moreover

$$\psi \in C_K \text{ if and only if } \sum_{t\in T} K_K(\psi,t) = \|K\|\ . \tag{3.6.7}$$

Using the above notations we can prove the following

Theorem III.6.3. Let $\mathcal{V} \subset \mathcal{K}(l_1,D)$ be a convex set and let $K \in \mathcal{K}(l_1,D)$, $V \in \mathcal{V}$. Then we have:

(a) $V \in P_{\mathcal{V}}(K)$ if and only if for every $U \in \mathcal{V}$ there exists $\psi \in C_{K-V}$ such that

$$re\Big(\sum_{t\in T} K_{U-V}(\psi,t)\cdot sgn(K_{K-V}(\psi,t))\Big) - \sum_{t\in A_\psi} |K_{U-V}(\psi,t)|\Big) \le 0. \qquad (3.6.8)$$

(b) V is a SUBA to K in $\mathcal{V}$ with a constant r>0 if and only if for every $U \in \mathcal{V}$ there exists $\psi \in C_{K-V}$ with

$$re\Big(\sum_{t\in T} K_{U-V}(\psi,t)\cdot sgn(K_{K-V}(\psi,t))\Big) - \sum_{t\in A_\psi} |K_{U-V}(\psi,t)|\Big) \le -r\cdot\|U-V\|, \quad (3.6.9)$$

where $A_\psi = \{t \in T: K_{K-V}(\psi,t)=0\}$.

Proof.a) Fix $K \in \mathcal{K}(l_1,Y)$ and $V \in P_{\mathcal{V}}(K)$. Let $K-V = \sum_{i=1}^{n} f_i(\cdot)\cdot y_i$. Assume that for some $U \in \mathcal{V}$ (3.6.8) is not fulfilled. Suppose $U-V = \sum_{i=1}^{n} g_i(\cdot)\cdot y_i$ and take $\phi \in E(K-V)$. We show that $re(\phi(U-V)) > 0$. To do this, we note that following Theorem (1.8) and Corollary (1.10) $\phi = \psi \otimes \gamma$, where $\psi \in ext\, S_{l_1^{**}}$ and $\gamma \in ext\, S_{l_1^*}$. By (3.1.28), we may assume that $\gamma \in S_{l_\infty}$ and $|\gamma(t)| = 1$ for every $t \in T$. Observe that

$$\|K-V\| = \phi(K-V) = \gamma((K-V)^*\psi) = \gamma\Big(\sum_{i=1}^{n}\psi(f_i)\cdot y_i\Big) = \sum_{t\in T}\gamma(t)\cdot K_{K-V}(\psi,t) \le$$

$$\le \sum_{t\in T}|K_{K-V}(\psi,t)| \le \|K-V\|.$$

By (3.6.7), $\psi \in C_{K-V}$. Hence $\gamma(t) = sgn\, K_{K-V}(\psi,t)$ if $t \in T \setminus A_\psi$. Compute

$$re(\phi(U-V)) = re\Big(\gamma\Big(\sum_{i=1}^{n}\psi(g_i)\cdot y_i\Big)\Big) = re\Big(\sum_{t\in T}\gamma(t)\cdot K_{U-V}(\psi,t)\Big) =$$

$$= re\Big(\sum_{t\in T} K_{U-V}(\psi,t)\cdot sgn(K_{K-V}(\psi,t))\Big) - \sum_{t\in A_\psi} K_{U-V}(\psi,t)\cdot(-\gamma(t))\Big) \ge$$

$$\ge re\Big(\sum_{t\in T} K_{U-V}(\psi,t)\cdot sgn(K_{K-V}(\psi,t))\Big) - \sum_{t\in A_\psi} |K_{U-V}(\psi,t)|\Big) > 0.$$

By Theorem (1.1), $V \notin P_{\mathcal{V}}(K)$.

Now suppose $V \notin P_{\mathcal{V}}(K)$ and take $U \in \mathcal{V}$ with $\|K-U\| < \|K-V\|$.

Choose $\psi \in C_{K-V}$ and define $\gamma \in ext(S(l_\infty))$ by

$$\gamma(t) = \begin{cases} sgn(K_{K-V}(\psi,t)) & ;K_{K-V}(\psi,t) \ne 0 \\ -sgn(K_{U-V}(\psi,t)) & ;K_{K-V}(\psi,t) = 0 \text{ and } K_{U-V}(\psi,t) \ne 0. \\ 1 & ;\text{in the opposite case} \end{cases}$$

Let $\phi = \psi \otimes \gamma$. Following ([118]), $\phi \in ext\, S_{\mathcal{K}^*(l_1)}$. Observe that, by Remark (1.7) and (3.6.7),

$$\phi(K-V) = \gamma(\sum_{i=1}^{n} \psi(f_i)\cdot y_i) = \sum_{t\in T} \gamma(t)\cdot K_{K-V}(\psi,t) = \sum_{t\in T\setminus A_\psi} |K_{K-V}(\psi,t)| =$$

$$= \|K-V\|.$$

Hence, by Theorem (1.1), $re(\phi(U-V)) > 0$. But

$$re(\phi(U-V)) = re(\sum_{t\in T} \gamma(t)\cdot K_{U-V}(\psi,t)) =$$

$$= re(\sum_{t\in T} K_{U-V}(\psi,t)\cdot \mathrm{sgn}\, K_{K-V}(\psi,t) - \sum_{t\in A_\psi} |K_{U-V}(\psi,t)|),$$

which gives the desired result.***

By (3.6.7), Theorem (4) of [45] and the similar reasoning as in Theorem (6.3), we can prove the following

Theorem III.6.4. Let $\mathcal{V} = \mathcal{P}(l_1, D)$ and $K = 0$. Assume furthermore that $\dim(Y|_A) = \dim Y$ for every infinite set $A \subset \{t\in T: y(t)\neq 0 \text{ for some } y \in D\}$.

Then we have:

(a) $V \in \mathcal{V}$ is a minimal projection if and only if for every $U \in \mathcal{V}$ there exists $s \in T$ with $\hat{s} \in C_{K-V}$ (compare with (3.6.7)) such that

$$re(\sum_{t\in T} K_{V-U}(\hat{s},t)\cdot \mathrm{sgn}(K_{K-V}(\hat{s},t)) - \sum_{t\in A_s} |K_{V-U}(\hat{s},t)|) \le 0;$$

(b) $V \in \mathcal{V}$ is a SUBA to K in $\mathcal{V}$ with the constant $r>0$ if and only if for every $U \in \mathcal{V}$ there exists $s \in T$ with $\hat{s} \in C_{K-V}$ such that

$$re(\sum_{t\in T} K_{V-U}(\hat{s},t)\cdot \mathrm{sgn}(K_{K-V}(\hat{s},t)) - \sum_{t\in A_s} |K_{V-U}(\hat{s},t)|) \le -r\cdot\|U-V\|.$$

Part a) of Theorem (6.4) in the real case has been proved (by a different method) in ([45],Th.5).

Example III.6.5. In $\mathbb{R}^4$ consider the four points $y_1 = (1,0,\alpha,\alpha)$, $y_2 = (0,1,-\alpha,\alpha)$, $u_1 = (1,-2\cdot\alpha\cdot\beta,0,\beta,\beta)$ and $u_2 = (0,1,-2\cdot\alpha\cdot\beta,-\beta,-\beta)$ in which α is a parameter satisfying $1/2 \le \alpha \le 1$ and $\beta = (\alpha - 1/2)\cdot (2\alpha^2- 2\alpha + 1)^{-1}$. One can verify that $(u_i,y_j) = \delta_{ij}$ (the symbol $(\cdot,\cdot)$ denotes the inner product in $\mathbb{R}^4$). Thus the maps $P = \sum_{i=1}^{2} u_i(\cdot)\cdot y_i$ and $P^* = \sum_{i=1}^{2} y_i(\cdot)\cdot u_i$ are projections. If we regard P as a projection of l_∞^4 and P^* as a projection of l_1^4, both are minimal. The function A_P (3.6.2) is constantly equal to $\alpha\cdot(2\alpha^2- 2\alpha + 1)^{-1}$ and the same is true for P^*. In veryfing the minimality, the cases $\alpha = 1/2$ and $\alpha = 1$ are done separately, for in these cases, $\|P\| = 1$. If $1/2 < \alpha < 1$ then the function $K_P(s,t)$ has the following matrix representation

$$\begin{bmatrix} \gamma & 0 & \beta & \beta \\ 0 & \gamma & -\beta & \beta \\ \alpha\gamma & -\alpha\gamma & 2\alpha\beta & 0 \\ \alpha\gamma & \alpha\gamma & 0 & 2\alpha\beta \end{bmatrix}$$

with $\gamma = 1 - 2\alpha\beta$. The three parameters α, β, γ are all positive. In verifying that P is minimal, in view of Theorem 6.1 a), it is enough to show that there do not exist vectors z_1, z_2 in $\mathbb{R}^4$ such that $(z_i, y_j) = 0$ and

$$\sum_{i=1}^{4} \sum_{i=1}^{2} z_i(t)\cdot y_i(s)\cdot \operatorname{sgn} K_P(s,t) > 0.$$

The inequalities for the z_i are these:

$$z_1(1) + z_1(3) + z_1(4) > 0$$

$$z_2(2) - z_2(3) + z_2(4) > 0$$

$$z_1(1) - z_2(1) - z_1(2) + z_2(2) + z_1(3) - z_2(3) > 0$$

$$z_1(1) + z_2(1) + z_1(2) + z_2(2) + z_1(4) + z_2(4) > 0.$$

Any vector z orthonormal to D must satisfy

$$z(1) = -\alpha\cdot z(3) - \alpha\cdot z(4) \text{ and}$$

$$z(2) = \alpha\cdot z(3) - \alpha\cdot z(4).$$

These two equations can be used to eliminate $z_i(1)$ and $z_i(2)$ from the system of inequalities, and the resulting system is easily seen to be inconsistent.

The minimality proof for P^* is almost the same, and we omit it.***

Notes and remarks

1. Theorem III.1.1 was proved by B. Brosowski and R. Wegmann in [30]. Theorem III.1.3 which developes Theorem III.1.1 was obtained by A. Wójcik in [189]. The crucial result of this section, Theorem III.1.6 was established by H.S. Collins and W. Ruess in [51].

2. The method of applying Theorem III.1.6 to approximation in spaces of compact operators has been noticed by the second author in [117]. The main results of this section Theorems III.2.2 and III.2.5 were proved by the second author in [117]. Theorem III.2.8 (in the real case) was obtained by E.W.Cheney and P.D. Morris in [41], but the method of the proof was different from the one used in section 2. Corollaries III.2.11 and III.2.12 were established in [41]. The way of proving Theorem II.3.6 presented in this section is due to the second author.

3. Theorem III.3.1 was proved by the second author in [117]. It is worth saying that in [12] M. Baronti gives a complete choracterization of these hyperplanes in l_∞ which are range of a SUBA projection. Proposition III.3.4 and the main result of this section Theorem III.3.11 was established by second author in [12]. In [12] there was also another proof of Theorem II.4.9 given without using mathematical programming method.

4. The results of this section have their origin in the following theorem due to E.W. Cheney [38]

Theorem III.4.5. Let $P_o \in I(C(T,\mathbb{R}),D)$. In order to $P_o \in \Delta(C(T,\mathbb{R}),D)$ it is necessary and sufficient that for every $P \in \mathcal{P}(C(T,\mathbb{R}),D)$ there exists $\hat{t} \in \text{crit}^* P_o$ such that $\inf\{((P-P_o)x)t : x \in A_t\} \le 0$.

Note that the problem of finding minimal projection in class $I(C(T,\mathbb{R}),D)$ is very difficult. For more details about this problem the reader is referred to [28] and [101].

5. Theorem III.5.1 was proved by the second author in [117]. Formula (3.5.6) was obtained, by a different method in [42]. Other results of this section were established by E.W. Cheney P.D. Morris and K.H. Price in [42].

6. The results of this section have their origin in the paper of E.W. Cheney and C. Franchetti [45]. Theorems III.6.1 and III.6.3 were established by the second author in [117]. In the real case Theorem III.6.1 a) and III.6.4 a) were proved by a different method in [45]. Example III.6.5 is due to E.W. Cheney and C. Franchetti [45].

Chapter IV

Isometries of Banach spaces and the problem of characterization of Hilbert spaces

§ 1. Isometries and minimal projections

In this section (as in the remaining ones, in fact) we use the following notation. If A is a isometry of a Banach space B onto itself, we write

$$B^{A} = \{x \in B:\ Ax = x\},\ B_{A} = (A-I)(B), \qquad (4.1.1)$$

I denoting the identity operator in B.

Theorem IV.1.1. ([107]) Let A be a linear isometric operator of the Banach space B onto itself, and $B = \mathrm{im}(I-A) \oplus \mathrm{Ker}(I-A)$. Let P_o be the projection from B onto $\mathrm{Im}(I-A)$ annihilated on $\mathrm{Ker}(I-A)$. Then P_o is a minimal projection and it can be defined by

$$P_o = \lim_{n\to\infty} \Big(\sum_{k=0}^{n-1} A^{-k}\circ P \circ A^{k}\Big)/n \qquad (4.1.2)$$

where P is a projection from B onto $\mathrm{Im}(I-A)$; $A^{o} = I$.

Proof. Let P be a projection onto subspace B_A. Let P_o be the map defined by (4.1.2).

We shall show that P_o is defined correctly for each $x \in B$ and it is a projection from B onto B_A along B^A. Indeed, if $z \in B_A$ then there exists a $x \in B$ such that $z = x - Ax$. Then $A^k z = ((I-A)\circ A^k)x \in B_A$ for all $k \in \mathbb{N}$. Hence $(P\circ A^k)z = A^k z$ for all $k \in \mathbb{N}$ and $P_o z = z$.

If $z \in B^A$, then $A^k z = z$ for all $k \in \mathbb{N}$. Since $Pz \in B_A$ and $Pz = x_1 - Ax_1$ for a $x_1 \in B$, we have

$$P_o z = \lim_{n\to\infty} (1/n)\cdot \sum_{k=0}^{n-1} A^{-k}(x_1 - Ax_1) = \lim_{n\to\infty} (1/n)\cdot(A^{-n+1}x_1 - Ax_1) = 0,$$

because $\|A^{-1}\| = 1$. By triangle inequality we obtain $\|P_o\| \le \|P\|$. Therefore, by linearity and boundedness of the map P_o, P_o is the minimal projection from B onto B_A along B^A.

Remark IV.1.2. The proof of this theorem is essentially a proof of some ergodic statistic theorem (cf.[60]). For more detailed references the reader is referred to [107].

For using of Theorem 1.1 we ought to have a condition for a decomposition of B as a direct sum $B = B_A \oplus B^A$. The next proposition gives the conditions of this decomposition (if A is a linear continuous map) in terms of a convergence of averaging operators A(n) defined by

$$A(n) = (1/n)\cdot\sum_{k=0}^{n-1} A^k \quad (n=1,2,\ldots) \tag{4.1.3}$$

Example IV.1.3. (Fürstenberg) Let $C(S_1,\mathbb{R})$ be the space of all continuous functions on the circle S_1. Then for every isometric surjective operator $A: C(S_1,\mathbb{R}) \to C(S_1,\mathbb{R})$ and for every $x \in S_1$ there exists $\lim_{n\to\infty} (A(n))x$, where A(n) is defined by (4.1.3).

Proposition IV.1.4. Suppose that a linear continuous operator A in a Banach space B is such that B_A is closed and

$$\lim_{n\to\infty} \|A(n)\|/n = 0. \tag{4.1.4}$$

In order that B be a direct sum $B_A \oplus B^A$ it is necessary and sufficient that

$$\text{there exists } \lim_{n\to\infty} (A(n))x \text{ for any } x \in B. \tag{4.1.5}$$

Proof. *Necessity.* Notice that if $x \in B^A$, then $(A(n))x = x$, and therefore $\lim_{n\to\infty} (A(n))x = x$. If $x \in B_A$, there exists $y \in B$ such that $x = y - Ay$, i.e., $(A(n))x = (1/n)\cdot(y - A^n y)$. Hence, by condition (4.1.4), $\lim_{n\to\infty} (A(n))x = 0$. Thus, for each $x \in B$, (4.1.5) is true.
The proof of sufficiency follows directly from [60] (Chapter VIII, § 5.2)***

Remark IV.1.5. (a) If A is an isometry, i.e., a linear isometric operator in B, then from the proof of necessity we have $B_A \cap B^A = \{0\}$. In general case the last equation is not true. For example, if $B = \mathrm{span}\{e^x, x\cdot e^x\} \subset C([0,1],\mathbb{R})$ and A is the differentiation operator in B, then $B^A = B_A = \mathrm{span}\{e^x\}$.
(b) If A is an isometry of B into itself and B is a reflexive space, then $B = \bar{B}_A \oplus B^A$ (cf.[60]), where $\bar{B}_A$ is the closure of $B_A = (I-A)B$.
(c) For the convergence of the sequence $(A(n))_{n=1}^{\infty}$ it is not sufficient that A can be an isometry of B onto itself. Indeed, let $\mathcal{A} = (a_{ij}), 1\le i,j\le 2$,

be a matrix of order two, $a_{ij} \in \mathbb{N}$, $1 \le i,j \le 2$, $\det \mathcal{A} = 1$, $|\mathrm{tr}\ \mathcal{A}| > 2$. (For example $\mathcal{A} = \begin{Bmatrix} 2 & 5 \\ 1 & 3 \end{Bmatrix}$.)

Let T be a standard automorphism of a torus $V = S_1 \times S_1$ corresponding to $\mathcal{A}$, i.e., for a point $v = (z,w) \in V$, $z = e^{2\pi i x}$, $w = e^{2\pi i y}$, $x,y \in \mathbb{R}$, we we have $Tv = (e^{2\pi i(a_{11}x+a_{12}y)}, e^{2\pi i(a_{21}x+a_{22}y)})$.

Let A be an isometry of the space $C(V,\mathbb{R})$ generated by T (i.e., $A\phi(v) = \phi(Tv)$, where $\phi \in C(V,\mathbb{R})$, $v \in V$). Then by a result of H. Fürstenberg from [193] (Theorem 3.3) in $C(V,\mathbb{R})$ there exists a function f for which the sequence $\{(A(n))f\}$ is not convergent, though A is a surjection. By Proposition 1.4, $C(V,\mathbb{R}) \ne (C(V))_A \oplus (C(V))^A$.

By the Banach Open Mapping Principle one can easy verify

Proposition IV.1.6. Let $A: B \to B$ be a linear continuous operator in B. Then the following statements are equivalent:

(i) $B = B_A \oplus B^A$;

(ii) B_A is a subspace of B and $(I-A)|_{B_A}$ is a one-one operator of B_A onto B_A;

(iii) for each $x \in B$ there exists $x_1 \in B$ such that $(I-A)x = (I-A)^2 x_1$ and for each sequence (x_k) such that $\lim_{k\to\infty} (I-A)^2 x_k = 0$, there holds $\lim_{k\to\infty} (I-A)x_k = 0$.

Remark IV.1.7. Let $A: B \to B$ be a linear continuous operator. It is easy see that

(a) If $A(B) = B$ and $B = B_A \oplus B^A$, then $A(B_A) = B_A$ and $B^{A^{-1}} = B^A$, $B_{A^{-1}} = B_A$, where A^{-1} is the inverse operator to A.

(b) If $\dim B_A < \infty$, then $B = B_A \oplus B^A$.

Example IV.1.8. Let M_i $(i=1,2)$ be two Orlicz functions: $[0,+\infty) \to [0,+\infty)$, i.e., continuous convex non-decreasing functions with $M_i(0) = 0$ and $M_i \ne 0$. Let $L_{M^2}(0,1)$ be an Orlicz space of equivalence classes of such measurable functions $h: [0,1] \to (-\infty,+\infty)$ for which

$$\|h\| = \inf\{t>0: \int_0^1 M_2(|h(x)|/t)\, dx \le 1\} < +\infty.$$

Let $B = L_{M^1}([0,1]; L_{M^2}(0,1))$ be the Orlicz space of the equivalence classes of strongly measurable functions $f: [0,1] \to L_{M^2}(0,1)$ for which

$$\|f\| = \inf\{t>0: \int_0^1 M_1(\|f(x)\|/t)\, dx \le 1\} < +\infty.$$

Now, we shall define for every $n \in \mathbb{N}$ a mop $\tau_n: [0,1] \to [0,1]$ as follows: if $n=1$, then $\tau_1(x) = x$ for every $x \in [0,1]$; if $n \geq 2$, then

$$\tau_n(x) = \begin{cases} x + (1/n) & \text{if } x \in (0,(n-1)/n); \\ x - (n-1)/n & \text{if } x \in [(n-1)/n,1]; \\ 1 & \text{if } x = 0. \end{cases}$$

An operator $Q(x): L_{M^2}(0,1) \to L_{M^2}(0,1)$ defined for every $x \in (0,1]$ by $(Q(x)h)y = h(\tau_{[1/x]}(y))$ (where $h \in L_{M^2}(0,1)$, $[1/x]$ is the greatest integer of the number $1/x, y \in [0,1]$), and $Q(0)h = h$, for $x = 0$, is a linear isometry.

Then the operator $A: B \to B$, such that $Af(x) = Q(f(\tau_2(x)))$ for each $f \in B$, $x \in [0,1]$ is a linear isometry of B onto itself.

B^A will be a subspace of all functions $f \in B$ satisfying for every $n \in \mathbb{N}$ the property

$(f(x))(\tau_n(y)) = (f(\tau_2(x)))(y)$ for μ-almost every $x \in (1/(n+1),1/n)$

and $y \in [0,1]$ relative to Lebesgue measure μ.

B_A will be a subspace of all functions $f \in B$ satisfying for every $n \in \mathbb{N} \setminus \{1\}$ the property

$$\sum_{k=0}^{n-1} (f(x)+f(x+(1/2)))\cdot(y+(k/n)) = 0 \tag{4.1.6}$$

for μ-almost every $x \in (1/(n+1),1/n]$ and $y \in [0,1/n]$ relative to Lebesgue measure μ.

It is easy to verify that the subspaces B^A and B_A are infinite-dimensional and the operator A satisfies condition (iii) from Proposition 1.6.

Now, let P be a projection from B onto B_A which can be defined in the following manner: $(Pf)x = f(x)$ if $x \in [0,1/2)$; if $n \geq 2$ and $x - (1/2) \in (1/(n+1),1/n]$, then

$$(Pf(x))y = \begin{cases} (f(x))y & \text{if } y \in [0,(n-1)/n); \\ -\sum_{k=0}^{n-1} (f(x-(1/2))+f(x))\cdot(y-(k/n)) & \text{if } y \in [(n-1)/n,1]. \end{cases}$$

By Theorem 1.1 and Proposition 1.6 the projection $P_0: B \to B_A$ defined by (4.1.2) is a minimal one. In virtue of the next theorem in this section $\|P\|_0 \le 2$.

Now we prove a theorem concerning evaluation of norm of minimal projections generated by isometry. At first we introduce some notions. Let D,K be subspaces of a Banach space B such that $B = D \oplus K$. Let $x \in S_k$ and

$D_x = D \oplus \mathrm{span}\,[x]$. Let $f \in S_{D_x^*}$ be such that $\ker f = D$. For every $a \in [0,1]$, let $W_a^x = f^{-1}(a)$, $C_a^x = W_{D_x} \cap W_a^x$, $\rho_D^x(a) = \inf\{\sup\{\|z-y\| : y \in C_a^x\} : z \in W_a^x\}$, C_a^x and $\rho_D^x(a)$ will be called respectively, the the hypercircle in D and the Chebyshev radius of C_a^x (cf.[65]).

Next write C_D^x for $\sup\{\rho_D^x(a) : a \in [0,1]\}$ and C_D^K for $\sup\{C_D^x : x \in S_K\}$. Consider now the function $g : [1,2] \to [1,2]$ (cf.[65]) defined as

$$g(t) = \begin{cases} 1+(1/2)\cdot((t-1)+((t-1)^2+8\cdot(t-1))^{1/2} & \text{if } 1 \le t \le \sqrt{17}-3; \\ 1+8(t-1)/(t^2+4(t-1)) & \text{if } \sqrt{17}-3 \le t \le 2. \end{cases} \tag{4.1.7}$$

Note that D is strictly increasing and concave. Moreover, $g(1) = 1$, $g(2) = 2$, $g(t) \ge t$ for each $t \in [1,2]$. In terms of the function g and the number C_D^K we can evaluate the norm of minimal projection P_o defined by (4.1.2).

Theorem IV.1.9. Let A be a linear isometry of Banach space B onto itself such that $B = B_A \oplus B^A$. Let $D = B_A$, $K = B^A$ and let $P_o \in \mathcal{P}(B,D)$ be a projection defined by (4.1.2). Then

$$1 \le C_D^K \le \rho(B,D) = \|P_o\| \le g(C_D^K) \le 2. \tag{4.1.8}$$

Proof. Let $x \in S_K$ and $D_x = D \oplus \mathrm{span}\,[x]$. By Remark 1.7 the operator $A_x = A|_{D_x}$ is an isometry of D_x onto itself with $\mathrm{Im}(I-A_x)=D$, $\ker(I-A_x)=\mathrm{span}\,[x]$. By Theorem 1.1, the projection $P_D^x : D_x \to D$ along span [x] is a minimal projection, i.e., $\|P_D^x\| = \rho(D_x,D)$. In view of the fact that the $\mathrm{codim}_{D_x} D = 1$ and by a result of C. Franchetti ([65],Th.3) we have

$$1 \le C_D^x \le \rho(D_x,D) \le g(C_D^x) \le 2.$$

Observe also that in view of the inequality $\rho(D_x,D) \le \rho(B,D)$, we obtain $1 \le C_D^K \le \rho(B,D)$. Next we use the fact that the projection P_o is minimal.

If $\|P_o\| = 1$, then the theorem follows from the identity $g(1) = 1$. Next assume that $\|P_o\| > 1$. Then for any $\varepsilon>0$ with $\varepsilon < \|P_o\|-1$, there exists $x_o \in S_B$ such that $\|P_o x_o\| + \varepsilon > \|P_o\|$. Let y_D and y_K be such that $x_o=y_D+y_K$ with $y_D \in D$ and $y_K \in K$. Evidently $y_K \ne 0$.

Let $z = y_K/\|y_K\|$. Then $\|P_o x_o\| = \|P_D^z x_o\| \le \|P_D^z\| = \rho(D_z,D)$, where $P_D^z \in \mathcal{P}(D_z,D)$ is the projection annihilated on span [z]. Clearly, $\rho(B,D) < \rho(D_z,D) + \varepsilon$. Since the function g is strictly increasing, we get

$\rho(B,D) \le \sup\{\rho(D_x,D): x \in S_K\} \le \sup\{g(C_D^x): x \in S_K\} \le$

$g(\sup\{C_D^x: x \in S_K\}) = g(C_K^D) \le 2.$ ***

By Theorem 1.9 and a result of C. Franchetti in [65] (Th. 4), taking into account the form of function g, we get directly:

Corollary IV.1.10. Let A be a linear isometry of a Banach space onto itself and let $B = B_A \oplus B^A$. Denote $B_A = D$ and $B^A = K$. Then

(I) $\rho(B,D) = 1$ iff $C_D^K = 1$ iff for every $x \in S_K$ $C_D^x = 1$ iff for every $a \in (0,1)$ and $x \in S_K$ $\rho_D^x(a) \le 1$;

(II) $\rho(B,D) < 2$ iff $C_D^K < 2$ iff for every $x \in S_K$ there exists $a_x \in (0,1)$ with $\rho_D^x(a_x) < 1 + a_x$;

(III) $\rho(B,D) = 2$ iff $C_D^K = 2$.

In the next example we give a realization of case (II) and (III) of Corollary 1.10. For case (I) see [124] (Proposition 3.a.4).

Example IV.1.11. Let B be a Banach space with the symmetric norm (relative to a normal basis $(e_i)_1^\infty$ ([167],[177])). Let $(j(i))_{i=1}^\infty$ be a strictly increasing sequence of natural numbers so that $j(1)=1$ and $k(i) = j(i+1)-j(i) \ge 3$. Now, let $A: B \to B$ be a linear isometry such that $Ae_s = e_{s+1}$ for every $s \in \mathbb{N}$ and $j(i) \le s \le j(i+1) - 1$ and $Ae_{j(i+1)-1} = e_{j(i)}$ $(i=1,2,\dots)$.

Let

$$D_{k(i)} = \{x=(\alpha_1,\alpha_2,\dots) \in B: \sum_{\nu=j(i)}^{j(i+1)-1} \alpha_\nu = 0,\ \alpha_\nu=0 \text{ if } (\nu<j(i) \text{ or } \nu\ge j(i+1))\},$$

$B_{k(i)} = \operatorname{span}[e_{j(i)},\dots,e_{j(i+1)-1}]$, $i=1,2,\dots$.

It is easy to see that $\dim B_{k(i)}/D_{k(i)} = 1$. From Theorem 1.1 it follows immediately that the isometry A generates in the subspace $B_{k(i)}$ the minimal projection $P_i \in \mathcal{P}(B_{k(i)},D_{k(i)})$.

Now, in view of Proposition 1.6 it is easy to check that its condition (ii) holds. Hence, by Theorem 1.1 the projection $P_o \in \mathcal{P}(B,D)$, where $D = \bigoplus_{i=1}^{\infty} D_{k(i)}$, along $\ker(I-A)$ is a minimal projection. It is obvious that

$$\|P_o\| = \sup\{\|P_i\|: i=1,2,\dots\} \qquad (4.1.9)$$

(because for each $x = \sum_{i=1}^{\infty} \alpha_i \cdot e_i$ we have $\|x\| \ge \|\sum_{\nu=j(i)}^{j(i+1)-1} \alpha_\nu \cdot e_\nu\|$, $i=1,2\dots$, [177]).

Now, let $B = l_1$ or $B = c_o$. We prove that $\|P_i\| = 2-2/k(i)$ $(i=1,2,\dots)$. Indeed, if $B = l_1$, then there exists a linear isometry $F_1: B_{k(i)} \to l_1^{k(i)}$, so $F_1(D_{k(i)}) = f^{-1}_{1,i}(0)$, where $f_{1,i} = (1,\dots,1) \in (l_1^{k(i)})^*$. If $B = c_o$, then there exists a linear isometry $F_2: B_{k(i)} \to l^{\infty}_{k(i)}$, so $F_2(D_{k(i)}) = f^{-1}_{2,i}(0)$, where $f_{2,i} = (1/k(i),\dots 1/k(i)) \in (l_{\infty}^{k(i)})^*$.

By the result of [45] we get in both cases: $\rho(B_{k(i)}, D_{k(i)}) = 2 - k(i)$. By (4.1.9), $\|P_o\| = 2 - \inf\{2/k(i): i=1,2,\dots\}$. Therefore, $\|P_o\| = 2$ and $C^K_D = 2$, where $K = \ker(I-A)$, if $\sup\{k(i); i=1,2,\dots\} = +\infty$. If $\sup\{k(i)\} < +\infty$, then for each $x \in S_K$ there exists $a_x \in (0,1)$ such that $\rho^x_D(a_x) < 1 + a_x$.

Remark IV.1.12. Note, if $B = c_o$, then the projection P_o from Example 1.11 is the unique minimal projection onto D.

If $B = l_1$, then the uniqueness of the minimal projection P_o onto D (from Example 1.11) fails, altough the projections P_i are unique minimal projections from $B_{k(i)}$ onto $D_{k(i)}$, $i=1,2,\dots$ (cf. [142], [155]).

From Theorem 1.9 we get immediately

Corollary IV.1.13. Suppose D be a complemented subspace of a Banach space B such that $\rho(B,D) > 2$. Then there exists no linear isometry A of B onto itself such that $D = \mathrm{Im}(I-A)$ and $B = D \oplus \ker(I-A)$.

Example IV.1.14. Let $B = C_o(2\pi)$ be the space of all continuous, 2π-periodic real-valued functions defined on $[0,2\pi]$, and D_n $(n\geq 1)$, be the subspace in B consisting of all trigonometric polynomials of degree $\leq n$.

We shall prove that for $n\geq 8$ there exists no linear isometry A of B into itself such that $\mathrm{Im}(I-A) = D_n$. Suppose that for some $n\geq 8$ there exists such a linear isometry A. By Proposition 1.6 and Example 1.3 it follows that that $B = D_n \oplus \mathrm{Ker}(I-A)$. Hence, by Theorem 1.9, $\rho(B,D_n) \leq 2$. On the other hand $\rho(B,D_n)$ is equal to the Lebesgue Constant ρ_n such that $\rho_n = (4/\pi^2)\cdot \ln(n) + 1.27033 + \varepsilon_n$, where $0.166 > e_n \geq 0$ (cf., for example [39]). Taking into account that $n\geq 8$, we have $\rho_n > 2$, a contradiction.

Remark IV.1.15. Note that in the last two examples, our results can also be obtained in a more complicated way with the help of the theory of operators acting on compact topological groups (cf., for example [124]).

Now, given a Banach space B let us denote by $\mathcal{G}_B$ the group of all isometries of B. If D is a subspace of B then

$$\mathcal{G}_{B,D} = \{A \in \mathcal{G}_B : (A-I)(B) = D \text{ and } B = D \oplus B^A\}. \tag{4.1.10}$$

Definition IV.1.16. Let D be a subspace of a Banach space B ($\dim D \geq 2$)

and suppose that $\mathcal{I}_{B,D} \neq \emptyset$. Let $E,F \in \mathcal{I}_{B,D}$. Isometries E and F are said to be D-equivalent iff $B^E = B^F$. Evidently this is an equivalence relation.

Theorem IV.1.17. Let D be a reflexive subspace of a Banach space B. Let A be an isometry of B into itself, $A \in \mathcal{I}_{B,D}$. Suppose that the minimal projection onto D is unique. Then all isometries in $\mathcal{I}_{B,D}$ are D-equivalent. Moreover, B^A is annihilated by this projection.

Proof. This is an immediate consequence of Theorem 1.1 and Definition 1.16. ***

Remark IV.1.18. a) Conditions which ensure the uniqueness of minimal projection onto D can be further specified; one can appeal e.g. to the results of Chapter I.

b) Sometimes Theorem 1.3 allows one to determine the minimal projection explicitly.

Example IV.1.19. Let B be either l_p, $2 \le p < \infty$, or c_o, or l_p^n, $2\le p<\infty$, $n\ge 3$. Let $\langle e_i \rangle$ be the canonical basis in B. Consider the subspace $D = [e_1 - e_2, -e_2 + e_3]$. Let σ denote the identity permutation of the set $\mathbb{N} \setminus \Gamma(3)$ or $\Gamma(n) \setminus \Gamma(3)$, according as dim B = ∞ or dim B = n. We define a linear operator in B by $Ae_1 = e_2$, $Ae_2 = e_3$, $Ae_3 = e_1$, $Ae_i = e_i$ for $i\ge 4$. Obviously, A is an isometry of B and we have $B^A = [e_1+e_2+e_3, e_4, e_5, \dots]$, $B_A = D$. Consequently, the projection onto D parallel to B^A is minimal. The uniqueness of this projection results by applying either Theorem I.2.13 (in case $B = l_p$, $2\le p<\infty$, or $B = l_p^n$, $2\le p\le\infty$, $n\ge 3$) or Theorem I.3.8 (in case $B = c_o$).

Let us also remark that, for $p>1$, the minimal projection considered in this example has norm strictly greater than one.

Remark IV.1.20. The converse of Theorem IV.1.3 is not true; i.e., the existence and uniqueness of a minimal projection onto a reflexive subspace D does not imply, in general, the existence of an isometry A of B onto itself such that $B_A = D$. Let us take, for instance, the three-dimensional space B whose unit ball is a right circular cylinder with bases not parallel to the plane D passing through the origin and perpendicular to the rotation axis. Clearly, there is no isometry A such that $D = B_A$, despite the fact that a minimal projection onto D exits and is unique.

Furthermore, a construction which we owe to W.J. Davis [56] shows that, starting with dimension three, there exist spaces admitting no isometries other that the trivial ones, I and $\mathcal{F}_I$ - the operator of central symmetry in B (hence no isometry A with $B_A \neq [0]$); yet, these spaces do contain subspaces D (even of codimension 1) with a unique minimal projection onto D.

Remark IV.1.21. The second part of this section deals mainly with isometries which generate minimal projections of unit norm. The problem of reproducing the generating isometry, given a norm 1 projection, has not been thoroughly studied. In this context, let us mention here the theorem proved by Yu.I.Ljubic ([125], Th.2) which asserts that eigenvectors corresponding to distinct eigenvalues of a subconservative operator in a complex Banach space are mutually orthogonal (Recall that a linear operator A in a complex Banach space B is called subconservative if $\|Ax - \lambda\cdot x\| > |\mathrm{im}\,\lambda|\cdot\|x\|$ for $x \in B$ and any complex λ.). In particular, if such an operator has only two eigenvalues then each of corresponding eigen-spaces is the image of a norm 1 projection from the direct sum of this subspaces.

Definition IV.1.22. Let A be an isometry of a Banach space B ($\dim B \geq 3$). A will be called an elementary rotation if

(i) $\dim B_A = 2$;

(ii) condition $A^2 = I$ implies that $A|_{B_A}$ coincides with $\mathcal{F}_o$, the operator of central symmetry in B_A (i.e. $A|_{B_A}$ is not a reflection);

(iii) $A^{2k+1} \neq I$, for any $k \in \mathbb{N}$.

Proposition IV.1.23. Let A be an isometry of a Banach space B ($\dim B \geq 3$) and suppose that $\dim B_A = 2$. The following conditions are (jointly) necessary and sufficient for A to be an elementary rotation:

a) B^A is the eigenspace corresponding to the eigenvalue $\lambda=1$ (of the operator A).

b) B_A is the eigenspace corresponding to the eigenvalue $\lambda=-1$ of the operator equal either to a power of A or to the limit of a sequence of powers of A.

Proof. The assertion of the proposition follows easily from the following fact ([147],Prop.1.2): if A is an isometry of a two-dimensional Banach space then either A^{2k+1} or A^{2k} is the identity (of the Banach space in question), for a certain $k \in \mathbb{N}$, or there exists an increasing sequence $\{k_n\}$ of positive integers such that $\lim A^{k_n}$ equals to $\mathcal{F}_I$, the operator of central symmetry in that space.***

Proposition IV.1.24. Suppose that A is an elementary rotation in B. Let P be the projection of B onto B_A parallel to B^A. Then $\|P\| = 1$.

Proof. Assume that $\|P\| > 1$. Then $\|Pz\| > 1$ for some $z \in S_B$. Fix z and write $x = Pz$, $y = z-x$. Clearly $y \in B^A$, $y \neq 0$. Let $D = B_A \oplus [y]$.

a) Suppose $A^2 = I$. According to Definition 1.8, $A_o = A|_{B_A}$ is the operator of central symmetry in B_A. Since $P \circ A = A \circ P$, $(P \circ A)z = (A \circ P)z = -x$. But $Az \in S_B \cap D$ and $Az \neq -z$ (else $z - Az \notin B_A$, contrary to the definition of B_A). The unit ball W_D is convex and hence contains the line segment $Q = [Az, -z]$. As $P(-z) = -x$, we get $x \in Q$ and thus $\|x\| \leq 1$, in contradiction to our assumption.

b) Now assume $A^2 \neq I$. Let $\delta_o = \inf\{\|x+v\| : v \in B_A \cap S_B\}$. Set $\varepsilon = \delta_o/(2 \cdot \|x_*\|)$, $x_* = x/\|x\|$. The fact which we used in the proof of Proposition 1.23 ([147],Prop.1.2) again apllies, showing that, for a certain $k \in \mathbb{N}$, $\|A^k x_* + x_*\| < \varepsilon$; or, which is the same, $\|A^k x + x\| < \delta_o/2$. Since $W_{B_A}(x;\delta_o/2) \cap S_B = \emptyset$ (by the definition of δ_o) the entire line segment $Q_1 = (A^k x, -x)$ lies outside W_D. Since $(P \circ A^k)z = A^k x$, the segment $Q_2 = [A^k z, -z]$ must intersect Q_1. But this is impossible, in view of the fact that the end-points of Q_2 belong to W_D which is a convex set.
This contradiction shows that $\|P\| = 1$.***

Remark IV.1.25. If condition (iii) is removed from Definition 1.22 then Proposition 1.10 does not hold any longer. To obtain a countrexample consider the situation of Example 1.19 taking $B = l^3_p$, $2<p\leq\infty$.

Now Proposition 1.24, Theorem 1.17 and Corollary I.2.20 combined result in

Corollary IV.1.26. Let D be a two-dimensional subspace of a Banach space B. Suppose that

(i) There is an elementary rotation which belongs to $\mathcal{I}_{B,D}$;

(ii) D^* contains a subset M^D, total in D, such that every functional in M^D has the property (U).

Then all isometries in $\mathcal{I}_{B,D}$ are D-equivalent and $B_A \perp B^A$.

Remark IV.1.27. Condition (ii) of the last corollary is certainly fulfilled if e.g. B is a smooth space. (For details see [144], Corollary 1).

Remark IV.1.28. Suppose A is an elementary rotation in B and P is a projection onto B_A parallel to B^A. Let $A_o = A|_{B_A}$ and write $I_o = I|_{B_A}$. Consider the quantity $c_A = \|A-I\| \cdot \|(A_o-I_o)^{-1}\|$. It can be shown that this quantity plays in B_A the role of a Lipshitz constant; i.e. for any $z \in B_A$ we have

$$\|A_o - I_o\| \le c_A \cdot \|Az - z\|. \qquad (4.1.11)$$

Further, it can be proved that $1 \le c_A \le 2$and, if $A^2 \ne I$, there is $n \in \mathbb{N}$ such that

$$1 \le c_A \le \cos^{-2}(\pi/2n) \qquad (4.1.12)$$

(the last bound being sharp).
Finally, the following estimations are true ([119], Corollary 2.2):

$$c_A = \max(\|A-I\|/\|y-A_o y\| : y \in B_A \cap S_B) \qquad (4.1.13)$$

$$\|A - I\| = c_A \cdot \|z - Az\|/\|Pz\| \quad \text{for } z \in S_B,\ z \notin B^A. \qquad (4.1.14)$$

The last inequality may be applied to introduce a certain characteristic (called the two-dimensional modulus of rotativity) of Banach spaces whose isometry groups contain at least one elementary rotation (see [147] for details).

§ 2. A characterization of inner product spaces within the class of uniformly smooth stricly normed Banach spaces

Definition IV.2.1. Let B be a Banach space over $\mathbb{R}$. Assume that to a given $x \in B \setminus \{0\}$ there corresponds a unique subspace D_x with $\dim B/D_x = 1$ so that if D is any other subspace of codimension 1 in B, $x \notin D$, then P_D^x, the projection onto D parallel to [x], has norm greater than $P_{D_x}^x$. We define a functional ϕ_x by setting for any $z \in B$, $z = a \cdot x + P_{D_x}^x z$,

$$\phi_x(z) = a \cdot \|x\| \quad (a \in \mathbb{R}) \qquad (4.2.1)$$

Lemma IV.2.2. Let B be a Banach space , let $x \in B \setminus \{0\}$, and consider the functional ϕ_x just defined. Then

a) ϕ_x is a linear functional,

$$\phi_{ax} = (\operatorname{sgn} a) \cdot \phi_x \quad \text{for } a \in \mathbb{R} \setminus \{0\} \qquad (4.2.2)$$

b) we have

$$\|\phi_x\| = \|I - P_{D_x}^x\|; \qquad (4.2.3)$$

c) the mutual inclinations of [x] to D_x and vice versa satisfy

$$([x]\hat{,}D_x) = \|\phi_x\|^{-1}, \tag{4.2.4}$$

$$(D_x\hat{,}[x]) = \|P_{D_x}\|^{-1}. \tag{4.2.5}$$

Proof. **a)** is obvious.

b) Accordingly to formula (2.2.1) and Definition 2.1,

$$P^x_{D_x} = I - \phi(\cdot)\cdot x. \tag{4.2.6}$$

Further, $\|\phi_x\| = \sup\{|\phi_x(z)| : z \in S_B\} = \sup\{\|z - P^x_{D_x} z\| : z \in S_B\} = \|I - P^x_{D_x}\|$.

c) follows directly from the definition of inclination (see [78], p.197) and formula (4.2.3).***

Lemma IV.2.3. Let B be a Banach space, $f \in B^*$, $f \neq 0$. Choose $x \in B \setminus D$, where $D = \ker f$. Let γ be the natural embedding of B into B^{**} and let $\hat{D} = \ker \gamma(x)$. Consider the projection $P^f_{D^\wedge} : B^* \to \hat{D}$ and $P^x_D : B \to D$. The following relations hold between their norms:

a) $\|P^f_{D^\wedge}\|_{B^*} \leq \|P^x_D\|_B$; (4.2.7)

b) if B is reflexive then

$$\|P^f_{D^\wedge}\|_{B^*} = \|P^x_D\|_B. \tag{4.2.8}$$

Proof. Let $f(x) = a \neq 0$. Set $f_o = a^{-1}\cdot f/\|f\|$. Clearly, $f_o(x)=1$ and $\gamma(x)f_o=1$. Then (see (2.1.1)) the two projections under consideration can be written in the form

$$P^x_D = I - f_o(\cdot)\cdot x, \tag{4.2.9}$$

$$P^f_{D^\wedge} = I^* - \gamma(x)(\cdot)\cdot f_o \tag{4.2.10}$$

and we have

$$\|P^f_{D^\wedge}\|_{B^*} = \sup\{\|v - ((\gamma(x)v)f_o\| : v \in S_{B^*}\} = \sup\{|v(z) - v(x)\cdot f_o(z)| : z \in S_B\}$$

$$= \sup\{|v(z - f_o(z)\cdot x)| : z \in S_B\} \leq \|v\|\cdot\sup\{\|z - f_o(z)\cdot x\| : z \in S_B\} = \|P^x_D\|_B,$$

proving (4.2.7).

b) Equality in (4.2.8) is an immediate consequence of (4.2.7) and the reflexivity of B.***

Theorem IV.2.4. Let B be a reflexive Banach space and suppose that every subspace $\hat{D} \subset B^*$ of codimension 1 admits a unique minimal projection from B^*. Then ϕ_x is well defined for every $x \in B \setminus \{0\}$.

Proof. Take an arbitrary $x \in B \setminus \{0\}$. Write $\hat{D} = \ker \gamma(x)$. By assumption, there is a unique minimal projection $P_{D^\wedge} : B^* \to \hat{D}$. Let $f \in (P_{D^\wedge})^{-1}(0)$, $\|f\| = 1$. Then $P_{D^\wedge} = P^f_{D^\wedge}$. Let $D = \ker f$. We verify that this D can be taken as the subspace D_x in Definition 2.1. Assume, contrary to the claim, that there exists a subspace $D_1 \neq D$ such that $\dim B/D_1 = 1$, $x \notin D_1$, $\|P^x_{D_1}\| \le \|P^x_D\|$. Let f_1 be a functional in B^* for which $D_1 = \ker f$. Obviously, $[f_1] \neq [f]$. By Lemma 2.3, $\|P^{f_1}_{D^\wedge}\| = \|P^x_{D_1}\|$ and $\|P^f_{D^\wedge}\| = \|P^x_D\|$, and hence $\|P^f_{D^\wedge}\| \ge \|P^{f_1}_{D^\wedge}\|$. Since the minimal projection onto $\hat{D}$ is unique, $P^f_{D^\wedge} = P^{f_1}_{D^\wedge}$, and this means that $f_1 \in [f]$, contrary to the choice of f_1.***

Corollary IV.2.5. Let B be a uniformly smooth space (i.e. a (UR)-space, see [58]) and suppose that B is also strictly normed. Then ϕ_x is well-defined for every $x \in B \setminus \{0\}$.

Proof. It follows from the uniform smoothness of B that B^* is uniformly convex and reflexive; and this implies the uniqueness of minimal projection in B^* (onto codimension 1 subspaces) having norm strictly greater that 1. (see Th.2.13). Since B is strictly normed, the smoothness of B^* is implied, and this in turn ensures the uniqueness of minimal projections in B^* with norm equal to 1 (see e.g. [141]).***

Remark IV.2.6. a) The smoothness of B is not a necessary condition for the existence of ϕ_x for all $x \in B \setminus \{0\}$.

Indeed, consider B^2, the two dimensional space with the norm induced by the convex set $\delta = \delta_1 \cap \delta_2$, where δ_1 and δ_2 are defined by the inequalities $x^2 + (y-4)^2 \le 25$ and $x^2 + (y+4)^2 \le 25$, respectively.

It is readily seen that for each $e \in B^2$ there exists a unique subspace D_e such that the projection onto D_e along $[e]$ has norm 1.

b) Also the reflexivity of B is not necessary, in general, for the existence of ϕ_x for a given $x \in B$, $x \neq 0$.

Let e.g. $B = c_o$. Let $x = (1/2, 1/2, 0, \ldots) \in c_o$. In virtue of Theorem 3 of [16], $\hat{D} = \ker \gamma(x)$ is a subspace of (c_o^*) admitting a unique minimal projection. $P_{D^\wedge}$ (its norm is equal to 1). Let $f \in P^{-1}_{D^\wedge}(0)$, $\|f\|_1 = 1$, and let $D = \ker f$. It is easy to show using Lemma 2.3 that D is just the subspace D_x of Definition 2.1; and it is well known that c_o is not reflexive.***

*

Lemma IV.2.7. Let H be a Hilbert space with norm induced by the inner product $\psi(\cdot,\cdot)$. Then we have for $x, y \in H$, $x \neq 0$,

$$\psi(x,y) = \|x\| \cdot \phi_x(y) \tag{4.2.11}$$

Proof. Note that, given any $x \in H \setminus \{0\}$, the space D_x in Definition 2.1 coincides with the space of all elements orthogonal to x. Therefore each element $y \in H$ is uniquely represented as the sum $y = a\cdot x + z$, where $z \in D_x$. Then $\psi(x,y) = \psi(x, a\cdot x+z) = a\cdot \|x\|^2 = \|x\|\cdot\phi_x(y)$. ***

Theorem IV.2.8. Let B be a uniformly smooth strictly normed Banach space. In order that B be isometrically isomorphic to a Hilbert space it is necessary and sufficient to require that

$$\|x\|\cdot\phi_x(y) = \|y\|\cdot\phi_y(x). \tag{4.2.12}$$

hold for every $x,y \in B \setminus \{0\}$.

Proof. Necessity follows from Lemma 2.7. We will prove sufficiency. We define a functional $\psi(\cdot,\cdot)$ on $B \times B$ by putting $\psi(x,y) = \|x\|\cdot\phi_x(y)$ if $x\neq 0$ and $\psi(0,y) = 0$. It is easy to show that taking into account the linearity of ϕ_x and equality (4.2.8) that $\psi(\cdot,\cdot)$ is a symmetric bilinear form. Finally, $\psi(x,x) = \|x\|^2$. Thus, we have introduced inner product in B which produces the norm. ***

Remark IV.2.9. Among all smooth Banach spaces, inner product spaces can be characterized by the condition (coincident with (4.2.12))

$$\|x\|\cdot g(x,y) = \|y\|\cdot g(y,x) \tag{4.2.13}$$

holding for all $x,y \neq 0$.

Moreover it follows from Lemma 2.7 that in a Hilbert space H the functional ϕ_x (of any point $x \in H \setminus \{0\}$) is identical with the Gâteaux differential at x; i.e.

$$\phi_x = g(x,\cdot). \tag{4.2.14}$$

The next proposition is an easy consequence of (4.2.2) and the fact that $\|g(x,\cdot)\| = 1$ for every $x \neq 0$ (see [164]):

Proposition IV.2.10. Let B be a uniformly smooth strictly normed space and let $x \in B \setminus \{0\}$. For the equality (4.2.14) to hold it is necessary and sufficient that

$$\|\phi_x\| = 1. \tag{4.2.15}$$

The last inequality can in general fail to hold, as could be seen from the following example.

Example IV.2.11. In the three-dimensional Euclidean space E^3 with coordinates (x,y,z) consider the convex body δ_1 resulting by rotation of a disc segment around its chord N_1N_2; it is assumed that the disc radius equals 100 and the endpoints of the chord are $N_1 = (0,0,50)$, $N_2 = (0,0,-50)$. Let $O_1 = (20,0,-50)$, $O_2 = (-20,0,50)$. Take two balls of radius $(6000)^{-1/2} - 2$ with centers in O_1 and O_2 and consider the intersection of δ_1 with these balls. Denote the resulting strictly convex body by δ_2. Let δ_3 be the envelope of δ_2 of radius 1. Now let us renorm E^3 with the Minkowski metric taken with respect to S^3, the boundary of δ_3. We obtain a space B, which is strictly normed and smooth (thanks to enveloping; see [187]). Let e be a unit vector in B parallel to the z-axis. It is not hard to see that the space D_e then consists of all vectors orthogonal (in E^3) to the z-axis, whereas the space $D = \{y \in B: g(e,y) = 0\}$ is parallel to the tangent plane to δ_3 at y_*, the point where the boundary S^3 is cut by z-axis.

It is a matter of a simple calculation to show that this plane is not parallel to plane z = 0, and hence $D_e \neq D$; i.e. $\phi_e \neq g(e,\cdot)$. But $\|\phi_e\|>1$. **

Remark IV.2.12. Given a uniformly smooth Banach space B and an element $x \neq 0$, let $\phi_x = g(x,\cdot)$ and let $P^x_{D_x}$ be the projection onto D_x along [x]. The norm of this projection does not at all need to be equal one. To see this consider the cube in E^3 of side length 20 situated symmetrically with respect to the coordinate axes. Again we take the radius one envelope of this cube and denote by B the space E^3 renormed in such a way that it has resulting convex body as its unit ball; B is a uniformly smooth space. Let v be the unit vector in B parallel to a diagonal of the original cube. Then it is easy to see that D_v is orthogonal (in E^3) to v and parallel to the tangent plane to the ball $W_B(0,1)$ at the point v; i.e., $\phi_v = g(v,\cdot)$. However, a simple calculation shows that $\|P^v_{D_v}\| > 1 + 1/30$. ***

Problem IV.2.13. Let B be a separable, uniformly smooth strictly normed Banach space , dim B ≥ 3. Suppose ϕ_x = const, for all $x \in B \setminus \{0\}$. Is B necessarily isometric and isomorphic to a Hilbert space?

Problem IV.2.14. Let B be a uniformly smooth strictly normed Banach space, dim B ≥ 3. Suppose $\phi_x = g(x,\cdot)$ for every $x \in B \setminus \{0\}$ (and hence $\|\phi_x\| = 1$). Must B be isometrically isomorphic to a Hilbert space?

As it may be seen from Theorems IV.3.9 and IV.3.12 of the next section (see also Remark IV.3.8), the requirement that B is separable is essential in the formulation of the Problem 2.13.

Remark IV.2 15. As regards the case of n=2, it is possible to construct a strictly normed smooth space B, dim B = 2, in which ϕ_x is a support functional for any $x \neq 0$ and yet B is not isometric to the Euclidean space. This goes as follows. Take a continuous function

$\tau:[0,\pi/2] \to [\pi/2,\pi]$, which is strictly increasing ($\tau(0)=\pi/2, \tau(\pi/2)=\pi$) and satisfies $\tau(\tau(\phi)) = \phi + \pi/2$. Define

$$\ln(r(\psi)) = \int_0^{\psi} \cos(\tau(\phi)-\phi)\, d\phi \text{ for } \psi \in [0,\pi),$$

$$r(\psi) = r(\psi-\pi) \text{ for } \psi \geq \pi.$$

The curve $\{(r(\psi),\psi): \psi \in [0,2\pi)\}$ (in polar coordinates) is the boundary of a two dimensional ball which defines a strictly convex and smooth space B. In this space, ϕ_x is a support functional, for each $x \neq 0$. And, of course one can without difficulty invent τ so that B be not isometric to E^2. This example is due to S.V.Konjagin.

Remark IV.2.16. In conclusion to this section, we ought to mention that the idea of describing the inner product spaces in terms of the properties of codimension one subspaces is not new at all. For instance, in P. Papini's paper (1982) ([157], Prop.2.2) we find the following result:

Let X be a normed space , dim $X \geq 3$, and let g^+ be the function on $X \times X \setminus \{0\} \times X$ defined by

$$g^+(x,y) = \lim_{t \to 0+} t^{-1}\cdot(\|x+t\cdot y\|-\|x\|).$$

For a subspace $Y \subset X$ with $\dim X/Y = 1$, let $\mathcal{P}_Y(x)$ denote the set of elements of best approximation in Y to a given element $x \in X$. Suppose that, whatever be a subspace $Y \subset X$, $\dim X/Y = 1$, the assumption: $x_o \in \mathcal{P}_Y(x)$ forces

$$g^+(x_o-y,x-y) \geq 0 \text{ for all } y \in Y. \tag{4.2.16}$$

Then X is an inner product space.

The result, in its turn, is relevant to the result of H. Berens and U. Westphal (1978) ([18],Th.3.1):

A space X is an inner product space if and only if, for any $Y \subset X$ with $\dim X/Y = 1$ and any $x,y \in X$, we have

$$g^+(x_o- y_o,x - y) \geq 0 \text{ for } x_o \in \mathcal{P}_Y(x),\ y_o \in \mathcal{P}_Y(y). \tag{4.2.17}$$

Remark IV.2.17. Let B be a three dimensional space defined in Remark 2.6.a). In that space, ϕ_x is a support functional not for all $x \in B \setminus \{0\}$, as it is easy to see. Nevertheless, the calss of support functionals coincides with $\{\phi_x : x \in S_B\}$. Let $\Lambda_B = \{\phi_x : x \in B\}$. If B is renormed into $E^2 = l_2^2$, then each $f \in \Lambda_B$ becomes a support functional in E^2.

Problem IV.2.18. Suppose that in a Banach space B the functional ϕ_x is defined for every $x \in B \setminus \{0\}$. Can one introduce a new norm in B so that all functionals ϕ_x would convert into support functionals?

§ 3. Properties of reflexive Banach spaces with a transitive norm

Lemma IV.3.1. Suppose B is a reflexive space. Let $\underline{\Delta}_1(B) = 1 + \alpha$ (see Section II.8). To every $x \in S_B$ there correcponds a subspace D_x, codim $D_x=1$, so that, given any other subspace $D \neq D_x$ with codim $D = 1$, we have

$$\|P^x_D\| \geq \|P^x_{D_x}\| \geq 1 + \alpha. \qquad (4.3.1)$$

Proof. Let $x \in S_B$. Consider the space $\hat{D}_x = \ker \gamma(x) \subset B^*$ (recall that γ denotes the natural embedding of B into B^{**}). Since B is reflexive, there exists a minimal projection $P_{\hat{D}_x}$ of B^* onto $\hat{D}_x$ (see [83],p.109). Let $f \in P^{-1}_{\hat{D}_x}(0) \cap S_{B^*}$, i.e. $P^f_{\hat{D}_x} = P_{\hat{D}_x}$. Let $D = \ker f$. According to Lemma 2.3 we get in view of condition (4.3.1) $\|P^x_D\| \geq 1 + \alpha$. Assume there exists $D_1 \subset B$ with codim $D_1 = 1$ and $\|P^x_{D_1}\| < \|P^x_D\|$. Consider the functional $f_1 \in S_{B^*}$ for which $D_1 = \ker f_1$. Then by Lemma 2.3 $\|P^{f_1}_{\hat{D}_x}\| = \|P^x_{D_1}\| < \|P^x_D\| = \|P^f_{D_x}\|$, contrary to the minimality of $P^f_{\hat{D}_x}$. This contradiction concludes the proof of (4.3.1).***

Remark IV.3.2. If B is a uniformly smooth space, dim $B \geq 3$, then D_x, the subspace from Lemma 3.1, is unique (see Theorem 2.4) and coincides with the subspace occuring in the definition of functional ϕ_k.

Theorem IV.3.3. Let B be an isotropic space (see Section II.8), dim $B = \infty$, and suppose that

$$\underline{\Delta}_1(B) = 1. \qquad (4.3.2)$$

Then B is isometrically isomorphic to a Hilbert space.

Proof. Choose an arbitrary three-dimensional subspace $B^3 \subset B$. Pick $x \in S_{B^3}$ and write $K = \ker \gamma(x) \subset (B^3)^*$. Take $\varepsilon>0$. There exists a subspace $D \subset B$ codim $D = 1$, such that $\rho(B,D) < 1 + \varepsilon$. Hence, there is a projection P_D of B onto D with $\|P_D\| < 1 + \varepsilon$. Let $y \in P^{-1}_D(0) \cap S_B$. Since B is isotropic, there exists an isometry $A: B \to B$ which carries y onto x. Write $D_1 = A(D) \cap B^3$. The operator $A \circ P^y_D \circ A^{-1}$ restricted to B^3 is a projection of B^3 onto D_1 and we have $\|P^x_{D_1}\| < 1 + \varepsilon$. Then by Lemma 4.1 ($\varepsilon$ being arbitrary) $(B^3)^*$ is projected onto K with norm one.

Since x was chosen arbitrarily, $(B^3)^*$ is isometric and isomorphic to the Hilbert space (i.e. to E^3) (see [7], [58]). In virtue of a Theorem of Frechet (see [7]) B^*, hence also B, is isometrically isomorphic to a Hilbert space .***

Remark IV.3.4. If B is isotropic and dim B < ∞, it is not necessary to impose condition (4.3.2) in order to assert that B is a Hilbert space.

Remark IV.3.5. As shown in [79] (see also [151]), every isotropic space which is either seperable or reflexive, is a smooth space.***

Proposition IV.3.6. Let B be a reflexive isotropic space, dim B ≤ ∞, and suppose that

$$\underline{\Delta}_1(B) = 1 + \alpha \quad (\alpha \geq 0). \tag{4.3.3}$$

For every $x \in S_B$ and subspace D_x (see Lemma 3.1) we then have

$$\|P^x_D\| = 1 + \alpha. \tag{4.3.4}$$

Proof. If either dim B < ∞ or $\alpha = 0$, then the assertion follows from Auerbach's result (see Remark 4.4) and from Theorem 4.3, respectively. Therefore we only need consider the case of $\alpha > 0$, dim B = ∞. Let $x \in S_B$. Choose arbitrary $\delta > 0$. In view of assumption (4.3.3) there exists a codimension 1 subspace $D_1 \subset B$ such that $\rho(B, D_1) < 1 + \alpha + \delta$. Let P_1 be a minimal projection onto D_1. Then $\|P_1\| < 1 + \alpha + \delta$. Take $y \in P_1^{-1}(0) \cap S_B$. Since B is isotropic, we can find an isometry A of B onto itself with $A(y) = x$. Then $A \circ P_1 \circ A^{-1}$ is a projection onto $A(D_1)$ of norm $< 1+\alpha+\delta$. Since $x \in (A \circ P_1 \circ A^{-1})^{-1}(0)$, $\|P^x_{D_x}\| < 1 + \alpha + \delta$; and since δ was arbitrary chosen, (4.3.4) is settled.***

Lemma IV.3.7. (Wojtaszczyk). Let B be a reflexive isotropic space. Then B^* is also isotropic.

Proof. At first recall that a point x in a closed set $D \subset B$ is said to be strongly exposed if, given any functional $f \in B^*$, the conditions $f(x) = \sup(f(y): y \in D)$ and $\lim_{n\to\infty} f(x_n) = f(x)$ (where $(x_n) \subset D$) imply $\lim_{n\to\infty} \|x_n - x\| = 0$. (see [20]). Now assume $f \in S_{B^*}$ is a strongly exposed point of the unit ball W_{B^*}; its existence is a consequence of reflexivity of B^* (see [124]). There is a point $x \in S_B$ such that

$$f(x) = \|x\| \cdot \|f\| = 1, \tag{4.3.5}$$

for which there exist no other functionals in S_{B^*} satisfying (4.3.5). Therefore, the ball W_B has a unique supporting functional at point x and so, B being isotropic, there is a unique support functional at every point of S_B. This means that B is a smooth space.
To prove the second part, given arbitrary $f, g \in S_{B^*}$, we can find elements $x, y \in S_B$ for which $f(x) = g(y) = 1$ holds. Let E be an isometry of B onto itself sending x into y; and let E^* be the operator on B^* defined by $(E^*h)(x) = h(Ex)$ for all $x \in B$, $h \in B^*$. Then $\|E^*g\| = \|g\| = 1$ and $(E^*g)(x) = g(Ex) = g(y) = 1$. Since the ball W_B has a unique support functional at x, we get $E^*g = f$, showing that B^* is an isotropic space.***

Remark IV.3.8. As it has been proved in [52] (see alse [121], Corollary 1), every reflexive Banach space B contains a dense set of points at which the norm of B is Gâteaux differentiable. Consequently, in a reflexive space, the boundary of the unit ball W_B certainly contains a point at which W_B has a tangent hyperplane.

The existence of a point on S_B with the unique supporting hyperplane can be also derived from the assumption of reflexivity (see [41], Th.5.9.8). Therefore, given that B is isotropic, the smoothness of B is easily hence deduced (see [188], [79]).

Theorem IV.3.9. Let B be a reflexive isotropic Banach space. Then:

a) B is smooth and strictly normed

b) Every codimension 1 subspace $D \subset B$ admits a minimal projection with norm equal to $1 + \alpha$, where $\alpha \geq 0$ depends on B only.

c) If B is uniformly smooth, then $\|\phi_x\| = \text{const}$, for all $x \in B \setminus \{0\}$.

Proof. a) By Lemma 4.7, the space B^* (as well as B) is smooth (see Remark 4.8) and hence, by duality ([58], [59]) B is strictly normed.
b) Let D be any subspace with $\dim B/D = 1$. Let $f \in S_{B^*}$ and $D = \ker f$. Since B^* is smooth and reflexive, f defines a subspace $\hat{D}_f$ as in Lemma 4.1. Let $\gamma(x)$ be an element in $S_{B^{**}}$ for which $\ker \gamma(x) = \hat{D}_f$. Then by Lemma 2.3 $\|P_D^x\| = \|P_{\hat{D}_f}^f\|$ and, by the definition of $\hat{D}_f$, $\|P_D^x\| = \rho(B^*, \hat{D}_f)$. Let $\Delta_1(B^*) = 1 + \mu$. Then Proposition 3.6 gives $\|P_{\hat{D}_f}^f\| = 1 + \mu$ and hence $1 + \alpha \leq \rho(B, D) \leq 1 + \mu$, so thar $\mu \geq \alpha$. The same reasoning applied to B^* and $B^{**} = \gamma(B)$ shows that $\mu \leq \alpha$. Thus $\mu = \alpha$, and so $\rho(B, D) = 1 + \alpha$.

c) Let $x,y \in S_B$. In virtue of Proposition 3.6, $\|P^x_{D_x}\| = 1 + \alpha = \|P^y_{D_y}\|$. Let A be an isometry of B onto B with $Ax = y$. The operator $A \circ P^x_{D_x} \circ A^{-1}$ is a projection onto $A(D_x)$, has norm $1 + \alpha$ and sends y into 0; thus, by the definition of D_y, we obtain: $D_y = A(D_x)$, $P^y_{D_y} = A \circ P^x_{D_x} \circ A^{-1}$.

Choose $z \in S_B$ and write $z_1 = A(z)$. Then (in view of Remark 3.2)

$$z = \phi_x(z) \cdot x + P^x_{D_x}(z), \text{ whence}$$

$$A(z) = \phi_x(z) \cdot A(x) + (A \circ P^x_{D_x})z, \text{ or } z_1 = \phi_y(z) \cdot y + P^y_{D_y} z_1.$$

On the other hand, $z_1 = \phi_y(z_1) \cdot y + P^y_{D_y} z_1$ and so $\phi_y(z_1) = \phi_x(z)$.

Now, A is an isometry and $A(S_B) = S_B$; thus $\|\phi_y\| = \|\phi_x\|$. To conclude the proof, it remains to appeal to Lemma 3.1.***

Remark IV.3.10. The proof of the last theorem shows that the following statements are valid for a reflexive isotropic space B, in which $\underline{\Delta}_1(B) = 1 + \alpha$;

1. For any subspace $D \subset B$, codim $D = 1$, we have

$$\rho(B,D) = 1 + \alpha \qquad (4.3.6)$$

2. To any $D \subset B$, codim $D = 1$, there exists $x \in S_B$ such that $D = D_x$.

3. If, in addition, B is uniformly smooth then, given any two subspaces D_1 and D_2 of codimension 1 in B, one can find an isometry A of B, which carries D_1 onto D_2.

4. For any subspace $\hat{D} \subset B^*$, codim $\hat{D} = 1$, we have

$$\rho(B^*,\hat{D}) = 1 + \alpha.$$

Proposition IV.3.11. (Wojtaszczyk; see [151]). Let B be a Banach (0,k)-space (see Introduction , Problem (1.0d)), dim $B \geq k+2 \geq 3$. Then B is isomerically isomorphic to a Hilbert space. (The case of k=1 is treated in [7], p.254).

Theorem IV.3.12. There exists an infinite-dimensional $(\alpha,1)$-space, $\alpha>0$, whose all subspaces of codimension 1 are mutually isometric, whereas the space itself is not isometric to a Hilbert space. Moreover, $\|\phi_x\| = \text{const}$ for all $x \neq 0$.

Proof. Let Ω be the product of a non-coutable set A by the closed interval [0,1]. Denote by Σ the σ-algebra of all sets $E \subseteq \Omega$ such that for each $a \in A$ the set $E_a = E \cap (\{a\} \times [0,1])$ is Lebesque measurable. Define a measure μ on Σ by the formula

$$\mu(E) = \sum_{\alpha\in A} |E_\alpha|,$$

$|\cdot|$ denoting the one dimensional Lebesque measure on $\langle a\rangle \times \mathbb{R}$. In virtue of a result of S. Rolewicz [140], $L_p(\Omega,\Sigma,\mu)$, where $1 < p < \infty$, is an isotropic reflexive space.***

Remark IV.3 13. It can be of interest to compare the last theorem with a result obtained in [185]: For every $n \geq 2$ there exists a non-Euclidean space of dimension n, which is isometric to its dual.

Problem IV.3.14. Does there exist a nonseperable Banach (α,k)-space with $\alpha > 0$, $K \geq 2$?

Problem IV.3.15. Does there exist a reflexive infinite-dimensional separable Banach (α,k)-space with $\alpha > 0$ and k satisfying $\dim B \geq k+2 \geq 3$?

A positive answer to the last question in the case k=1 was given by S. Rolewicz [168].

Theorem IV.3.16. (S.Rolewicz). Let $(X,\|\cdot\|)$ be the product of two Banach spaces $(X_1,\|\cdot\|^1)$ and $(X_2,\|\cdot\|^2)$. Assume that $\|(x_1,x_2)\| = f(\|x_1\|^1,\|x_2\|^2)$, where f is increasing in both arguments. Then, if X is an (α,k)-space, so is X_1.

Proof. Let Y be any subspace of X_1 of codimension k. Then $Y_1 = Y \times X_2$ is a codimension k subspace of X. Thus there is a projection P of X onto Y_1 of norm $\|P\| = 1 + \alpha$. Let P_1 denote the projection P restricted to X_1: of course, $\|P_1\| < 1 + \alpha$.

On the other hand, there is no projection Q of X_1 onto Y of norm less than $1+\alpha$. Indeed; suppose that Q is such a projection. We define a projection Q_1 of X onto Y_1 by $Q_1(x_1,x_2) = (Qx_1,x_2)$. According to the assumption concerning the norm in X,

$$\|Q_1(x_1,x_2)\| = \|(Qx_1,x_2)\| = f(\|Qx_1\|^1,\|x_2\|^2) \leq f(\|Q\|\cdot\|x\|^1,\|x\|^2) <$$
$$< (1+\alpha)\cdot f(\|x\|^1,\|x\|^2) = (1+\alpha)\cdot\|(x_1,x_2)\|.$$

But this contradicts the assumption that X is an (α,k)-space. Therefore, $\|P_1\| = 1 + \alpha$ and there is no projection of X_1 onto Y of smaller norm. Since Y was taken to be an arbitrary subspace of X_1, of codimension k, X_1 is an (α,k)-space.***

As a consequence we obtain

Theorem IV.3.17. (S.Rolewicz). $L_p[0,1]$ is an $(\alpha,1)$ space for $1 < p < \infty$, $p \neq 2$ $(\alpha > 0)$.

Proof. Consider $X = L_p(\Omega,\Sigma,\mu)$, the space defined in the proof of Theorem 4.12. Let $\Omega_1 = \{a\} \times [0,1]$, $\Omega_2 = (A\setminus\{a\}) \times [0,1]$ and let Σ_1, Σ_2, μ_1, μ_2 be the induced σ-algebras and measures. Obviously, $L_p(\Omega_1,\Sigma_1,\mu_1)$ is isometric to $L_p[0,1]$.

Let $\|\cdot\|^1$, $\|\cdot\|^2$ be the L_p-norms on $L_p(\Omega_1,\Sigma_1,\mu_1)$ and $L_p(\Omega_2,\Sigma_2,\mu_2)$. Observe that $\|(x_1,x_2)\| = (\|x_1\|^1 + \|x_2\|^2)^{1/p}$.

Thus by theorem 3.12 and 3.16, $L_p[0,1]$ is an $(\alpha,1)$-space.***

Notes and remarks

1. The results of this section were inspired by discussion with M.I. Kadec in 1976-77; the initial stimulation came from W. Wojtyński's paper [188] (despite a mistake in one the proof in it). The material concerning isometries and the generated minimal projections with unit norm has been partially included in [147] (1979) and partially (the results given in the first subsection) first published in [156]. Certain version of Theorem IV.1.1 has apparently been known for quite a time already. The Example IV.1.8 is due to A.L. Koldobski (see [107]). Compare this theorem with a theorem of Rudin ([48], Th.9.1).

2. The contents basically follows article [154] (1985). The example presented in Remark IV.2.15 is due to S.V. Konjagin.

3. Theorems IV.3.3 and Proposition IV.3.6 were obtained in 1980. Lemma IV.3.7 was communicated to the author by P. Wojtaszczyk in 1981. These results were published in [151] (1982). Problem IV.3.15 was posed also in that paper; the positive answer (for k=1) obtained by S. Rolewicz [168] (1986) is presented in Theorems IV.3.16 and IV.3.17.

References

[1] Yu. A. Abramovič, *Symmetric spaces*, (Russian), Functional. Anal. i Prilozen. 9 (1975), 45-46 = <Functional. Anal. Appl. 9, no.1 (1975) , 51-52 (English translation)>.MR 51 #3857.

[2] A. Aleksiewicz, *Functional analysis*, (Polish) Polish Scientific Publisher, vol 49 (1969)

[3] D. Amir, C. Franchetti, *A note on characterizations of Hilbert spaces*, Bolletino U.M.I. 2A (1983), 305-309.

[4] E. Asplund, *Averaged norms*, Israel J. Math. 5, No.4 (1967), 227-233. MR 36 #5660.

[5] H. Auerbach, *Sur le groupes linéaires*, Stud. Math. I No. 4 (1935), 113-127; II No. 4 (1935), 158- 166; III No. 5 (1936) 43-49.

[6] H. Auerbach, S. Mazur and S. Ulam, *Sur le proprieté de l'ellipsoide*, Monatshefte fur Math. und Phys., 42, (1935), 45-48.

[7] S. Banach, *Oeuvres*, , vol. 2, Polish Scientific Publishers, Warsaw, 1979.

[8] S. Banach, *Kurs functionalnego analiza* (Ukrainian) Radianskaja skola, Kiev, 1948.

[9] V.F. Babenko S.A. Pricugov, *On property of compact operators on the space of integrable functions*, Ukrain. Math. Zhur. 33 (1981), 491-492 (Russian).

[10] M. Baronti, P.L. Papini, *Norm one projections onto subspaces of* l^p, Ann. Mat. Pura Appl. IV (1988), 53-61.

[11] M. Baronti, C. Franchetti, *Minimal and polar projections onto hyperplanes in the spaces* l_p *and* l_∞, preprint.

[12] M. Baronti, G. Lewicki, *Strong unicity of minimal projections onto hyperplanes of* l_∞ *and* l_1^n, preprint.

[13] E.F. Beckenbach, R. Bellman, *Inequalities*, ed. Neue Folge-Heft 30, Springer-Verlag, Berlin, 1961).

[14] P.K. Belobrov, *The operator of minimal extention*, (Russian) Math. Zamietki, 21 No. 4 (1975), 539-550. ZB1.[9]

[15] P.K. Belobrov, *Minimal extention of linear functionals onto the second conjugate space*, (Russian) Mat. Zamietki 27 No. 3 (1980), 439-445. = < Math. Notes 27 No. 34 (1980), 218-221 (English transformation)> MR 81 m; 46028.

[16] L.P. Belluce, W.A. Kirk and E.F. Steiner, *Normal structure in Banach spaces*, Pacif. J. Math. 26 No. 3 (1968), 433-440. MR 38 #1501.

[17] J. Bergh and J. Löfström, *Interpolation spaces. An introduction*, Grundlehren der Mathematischen Wissenschaften, 223 Springer-Verlag, 1976. MR 58 #2349.

[18] H. Berens and U. Westphal, *Kodissipative metrische Projektionen in normietren linearen Räumen*. (English summary), in: Linear spaces and approximation , P.L. Butzer and B.Sz. Nagy, ed. I.S.N.M., vol 40, Birkhäuser, Basel, 1978, MR 58 #23521.

[19] C. Bessaga and A. Pełczyński, *On bases and unconditional convergence of series in Banach spaces*, Studia Math. 17 No. 1 (1958), 151-164.

[20] C. Bessaga and A. Pełczyński, *Selected topics in infinite dimensional topology*, Monografie matematyczne, 58, Polish Scientific Publishers, Warsaw, 1975, Mr 57 #17657.

[21] J. Blatter and E.W. Cheney, *Minimal projections onto hyperplanes in sequence spaces*, Ann. Mat. Pura Appl. 101 (1974), 215-227, MR 50 #10644.

[22] J. Blatter and E.W. Cheney, *On the Existence of Extremal Projection*, J. Approx. Th., 6 (1972), 72-79. MR 49 #3403.

[23] B. Beauzamy, *Points minimaux dans les espaces de Banach*, C.R.A.S., 280 (1975), 717-720. MR 51 #11066.

[24] B. Beauzamy and B. Mourey, *Points minmaux et ensamble optimaux dans les espaces de Banach*, J. Functional Analysis, 24 No. 2, (1977), 107-139. MR 55 #1044.

[25] H.F. Bohnenblust, *Convex regions and projections in Minkowski spaces*, Ann. of Math. 39 (1938), 301-308.

[26] H.F. Bohnenblust, S. Karlin and L. Shapley, *Solutions of discrete two-persons games*, *in:* Contributions to the Theory of Games, Annales of Mathematics Study 24, Princeton 1950, 51-72.

[27] F. Bonsall, *Dual extremum problems in the theory of functions*, J. London Math. Soc. 31 (1956), 105-110.

[28] C. de Boor, *Proof of the Conjectures of Bernstein and Erdös Concerning the Optimal Nodes for polynomial Interpolation*, J. Approx. Th. 24 (1978) 289-303.

[29] M.Š. Bravermann and E.M. Semenov, *Isometries of symmetric spaces*, (Russian) Dokl. Acad. Nauk SSSR, 217 No. 2 (1974), 257-259. = {Soviet Math. Dokl. 15 (1974), 1027-1029 (English translation)}.MR 50 #28.

[30] B. Brosowski and R. Wegmann, *Characterisierung bester Aproximationen in normierten Vektorraumen*, J. Approx. Th. 3 (1970), 369-397.

[31] V.A. Bulovski, M.A. Jakovlieva and R.A. Zviagina, *Numerical methods of linear programming (selected problems)*, (Russian) Library of Economics and Mathematics, Nauka, Moscow, 1977. MR 80 c: 90093.

[32] Yu.D. Burago and V.A. Zalgaller, *Geometric inequalities*, (Russian) Nauka, Leningrad 1980. MR. 82 d:52009.

[33] B.L. Chalmers ant F.T. Metcalf, *A chracterization and equations for minimal extentions and extentions*, preprint.

[34] B.L. Chalmers and F.T. Metcalf, *The determination of minimal projections and extentions in L_1*, preprint.

[35] E.W.Cheney, *Introduction to approximation theory*, Mc Graw-Hill, New York, 1966.

[36] E.W. Cheney, C.R. Hobby, P.D. Morris, F.Schurer and D.E. Wulbert, *On the minimal property of the Fourier projections*, Bull. Amer. Math. Soc. 75 (1969), 51-52. MR 38 #4833.

[37] E.W. Cheney, C.R. Hobby, P.D. Morris, F. Schurer and D.E. Wulbert, *On the minimal property of the Fourier projections*, Trans. Amer. Math. Soc. 143 (1969), 249-258. MR 41 #704.

[38] E.W.Cheney, *Minimal interpolating projections*, Internat. Ser. Numer. Math., vol. 15, Basel 1970, 115-121.

[39] E.W. Cheney and K.H. Price, *Minimal projections*, in: Approximation Theory, Proc. Symp. Lancaster, July 1969, ed. A. Talbot, London 1970, 261-289. MR 42 #751.

[40] E.W.Cheney, *Projections with finite carriers*, Proceedings of a Conference at Oberwolfach, June 1971, ISNM 16 (1972), 19-32.

[41] E.W. Cheney and P.D. Morris, *On the existsnce and characterization of minimal projections*, J. Reine Angew. Math. (270), (1974), 61-76.

[42] E.W. Cheney, P.D. Morris and K.H. Price, *On approximation operator of de la Vallè Poussin*, J. Approx. Th. 13, (1975), 375-391.

[43] E.W.Cheney, *A survey of recent progress in approximation theory*, in: Proceedings of the International Congress of Mathematicians, (Vancouver, B.C. 1974), vol. 2, 411-415, Canad. Math. Congress., Montreal, Que., 1975. MR 55 #929.

[44] E.W. Cheney and C. Franchetti, *Minimal projections in L_1 spaces*, Duke Math. J. 43, No. 3 (1976), 501-510, MR 54 #11044.

[45] E.W. Cheney and C. Franchetti, *Minimal projections of finite rank in sequence spaces*, Colloq. Math. Soc. Janos Bolyai, 9 (1976),241-243.

[46] E.W. Cheney and C. Franchetti, *Orthogonal projections in the space of continuous functions*, J. Math. Anal. Appl. 63 No. 1 (1978), 253-264, MR 58 #1923.

[47] E.W.Cheney and C. Franchetti, *Minimal Projections in Tensor Product Spaces*, J. Approx. Th. 41, (1984), 367-381.

[48] E.W. Cheney and W.A. Light, *Approximation theory in Tensor Product Spaces*, Lecture Notes in Math., No. 1169, Springer, (ed. A. Dold and B. Eckman), Berlin 1985.

[49] Z. Ciesielski, *The C(I) norms of orthogonal projections onto subspaces of polygonals*, Trudy Mat. Inst. Steklov. 134, (1975), 366-369. MR 52 #6277.

[50] H.B. Cohen and F.E. Sullivan, *Projecting onto cycles in smooth reflexive Banach spaces*, Pacif. J. of Math. 34, No. 2, (1970), 355-364. MR 42 #2283.

[51] H.S. Collins and W. Ruess, *Weak compactness in spaces of compact operators and vector valued functions*, Pacific J. Math. 106 (1983) 45-71.

[52] H.H. Corson and J. Lindenstrauss, *On weakly compact subsets of Banach spaces*, Proc. Amer. Math. Soc. 17 (1966) 407-412. MR 42 #7812.

[53] G.B. Dantzig, *Linear Programming and Extentions*, Princeton University Press, Princeton, 1963.

[54] I.K. Daygaviet, *A property of compact operators in the space C*, (Russian) Uspehi Matem. Nauk, 18, No. 5 (1963), 157-158. MR 28 #461.

[55] I.K. Daugawiet, *The finite dimensional projection operators in the space C, that have norm one*, (Russian), Mat. Zamietki, 27 No. 2 (1980), 267-272, = (Math. Notes, 27. No. 1-2 (1980) 132-134, (English translation)). MR 83 c:46026.

[56] W.J. Davis, *Seperable Banach spaces with only trivial isometries*, Rev. Roumain Math. Pur. et Appl. 16, No. 7 (1971), 1051-1054. MR 45 #7445.

[57] M.M. Day, *Reflexive Banach spaces not isomorphic to uniformly convex spaces*, Bull. Amer. Math. Soc.,47 (1941), 313-317.

[58] M.M. Day, *Normed linear spaces*, Springer-Verlag, Berlin 1958.

[59] J. Diestel, *Geometry of Banach spaces*, Lectures Notes in Math., vol 485, Springer-Verlag, New York, 1975.

[60] N. Dunford and J.T. Schwartz, *Linear Operators Part I*, Interscience Publishers, New York, London, 1959.

[61] A. Dvoretzky, *A theorem on convex bodies and applications to normed linear spaces*, Proc. Nat. Acad. Sci. USA, 45 (1959), 223-226.

[62] T. Figiel, *Some remarks on Dvoretzky's theorem on almost spherical sections of convex bodies*, Colloq. Math. 24 (1972), 241-252, MR 46 #8044.

[63] R.J. Fleming , J.A. Goldstein and J.E. Jamison, *One parameter groups of isometries on certain Banach spaces*, Pacif. J. Math. 64 No. 1 (1976), 145-151, MR 54 #3372.

[64] F. Forelly, *The isometries of H^p*, Canad. Math. J. 16 No. 4 (1964), 721-728. MR 29 #6336.

[65] C. Franchetti, *Projections onto Hyperplanes in Banach Spaces*, J. Approx. Th. 38, (1983), 319-333.

[66] C. Franchetti, *Approximation with Subspaces of Finite Codimension*, in: Complex Analysis, Functional Analysis and Approximation Theory, ed. J. Mujica, Elsevier Science Publisher, 1986.

[67] C. Franchetti, *A numerical evaluation of projection constants*, Linear Algebra and its Applications 109, (1988), 179-196.

[68] P. Franck, *Sur la plus courte distance d'un operateur aux operateurs dont le noyeau continent un ensamble donné ou est de dimension moins égale à un nombre donné*, C.R.A.S. 263, No. 12 (1966), 388-389. MR 34 #4904.

[69] D. Gale and S. Sherman, *Solutions of finite two-person games*, in: Contributions to the theory of games, Ann. of Mathematics Study, vol. 24, H.W. Kuhn and A.W. Tucker, eds., Princeton University Press, princeton, (1950), 37-49.

[70] D. Gale, H.W. Kuhn and A.W. Tucker, *Reduction of game matrices*, in: Contributions to the theory of games, Ann. of Mathematics Study, vol. 24, H.W. Kuhn and A.W. Tucker, eds., Princeton University Press, Princeton 1950, 89-96.

[71] A.L. Garkavi, *Duality theorems for approximation by elements of convex sets*, (Russian) Uspehi Mat. Nauk., 16 No. 4 (1961), 141-145. MR 24 #2828.

[72] A.L. Garkavi, *The theory of best approximation in normed linear spaces*, in: Matematičeskii Analiz, 1967, VINITI, Moscow, 1967, 75-132. = (Progress in Mathematics, vol. 8: Mathematical Analysis, Plenum, New York, 1970, 83-150 (English translation)). MR 43 #7843.

[73] G. Godini, *On minimal points*, Commentat. Mathem. Universit. Carolinae 21 No.3 (1980), 407-419. MR 81 j:46023.

[74] G. Godini, *On generalized minimal points*, in: Proceedings of the Fourth Conference on Operator Theory, Univ. Timişoara, (1980), 239-245. MR 83 f:46020.

[75] G. Godini, *Some remarks on minimal points in normed linear spaces*, Ann. Numér. et Theor. Approximat. 10 No. 1 (1981), 17-22.

[76] M.L. Gromov, *On a geometric hypothesis of Banach*, (Russian) Izv. Acad. Nauk SSSR, Ser. mat., 31 (1967), 1105-1114. MR 36 #655.

[77] B. Grünbaum, *Some applications of expansion constants*, Pacif. J. of Math. 10 No. 1 (1960), 193-201.

[78] V.I. Gurarii, *Openings and inclinations of subspaces of Banach space*, (Russian), Theor. Funkcii, Funkcional. Anal. i Priložen. Vyp. 1, (1965), 194-204. MR 33 #7816.

[79] V.I. Gurarii, *Spaces of universal placement, isotropic spaces and a problem of Mazur on rotations of Banach spaces*, (Russian), Sibirsk. Mat. J., 7 No. 5 (1966), 1002-1013. MR 34 #585.

[80] R.R. Holmes, *Geometric functional analysis and its applications*, F.W. Gehring, P.R. Halmos and C.C. Moore, eds. Graduate Texts in Mathematics 24, Springer-Verlag, New York, 1975.

[81] H. Hudzik, *Uniformly convex Musielak-Orlicz spaces with Luxemburg's norm*, Comment. Math. Prace Mat., 23 No. 1 (1983), 21-32.

[82] H. Hudzik, *Convexity in Musielak-Orlicz spaces*, Hokkaido Math. J., 14 No. 1 (1985), 85-96.

[83] J.R. Isbell and Z. Semadeni, *Projection constants and spaces of continuous functions*, Trans. Amer. Math. Soc. 107, No. 1 (1963), 38-48. MR 26 #4169.

[84] R.C. James, *Orthogonality and linear functionals in normed linear spaces*, Trans. Amer. Math. Soc. 61 No. 2 (1947), 265-292. MR 6-273.

[85] R.C. James, *Characterizations of reflexivity*, Stud. Math. 23 No. 3 (1964), 205-216. MR 30 #431.

[86] G.J.O. Jameson and A. Pinkus, *Positive and Minimal Projections in Functions Spaces*, J. Approx. Th. 37 (1983), 182-195. MR 84 f:46031.

[87] W.B. Johnson, B. Mourey, G. Schechtman and L. Tzafriri, *Symmetric structures in Banach spaces*, Memoirs of the Amer. Math. Soc., vol. 19 No. 217, 1979. MR 82 j:46025.

[88] M.I. Kadec, *Geometry of normed spaces*, (Russian) in: Mathematical Analysis, vol 13, VINITI, Moscow, 1975, 99-127. MR 58 #30064.

[89] M.I. Kadec and B.S. Mitjagin, *Complemented subspaces in Banach spaces*, (Russian) Uspehi Mat. Nauk 28 No. 6 (1973), 77-94 = (Math. Surveys, 28 No. 6 (1973), 77-95. (English translation)) MR 53 #3649.

[90] M.I. Kadec and M.G. Snobar, *Certain functionals on the Minkowski compactum*, (Russian) Matem. Zamietki 10 No. 4 (1971), 453-458 = (Math. Notes 10 (1971), 694-696 (English translation)). MR 45 #861.

[91] M.I. Kadec and A. Pełczyński, *Bases, lacunary sequences and complemented subspaces in the spaces* L_p, Studia Math. 21 No. 2 (1962), 161-176. MR 27 #2851.

[92] A. Kamińska, *On uniformly Orlicz spaces*, Indagationes Math. 44 No. 1 (1982), 27-36.
[93] L.V. Kantorovič, *Mathematical methods in the organization and planning of production*, Leningrad University, Leningrad, 1939.
[94] L.V. Kantorovič, *On the translocation of masses*, (Russian) Doklady Akad. Nauk SSSR, 37 No. 7 (1942), 227-229. MR 5-174.
[95] V.L. Kantorovič, *Economic calculation of optimal utilization of resources*, (Russian), Izd. Akad. Nauk SSSR, Moscow 1959.
[96] L.V. Kantorovič and G.P. Akilov, *Functional analysis*, (Second edition), (Russian) Nauka, Moscow 1977. MR 58 #23465.
[97] L.V. Kantorovič and M.K. Gavurin, *Application of mathematical methods in problems of analysis of loadflows*, (Russian) in: Problems of increasing the efficiency level of transport, Akad. Nauk SSSR, Moscow, 1949, 110-135.
[98] L.V. Kantorovič and G. Sh. Rubinstein, *On a certain function space and a certain extremum problems*, (Russian) Doklady Akad. Nauk SSSR, 115, No. 6 (1957), 1058-1061. MR 20 #1219.
[99] L.V. Kantorovič and V.A. Zalgaller, *Efficient cutting of industrial materials*, (Russian) Nauka, Novosibirsk, 1971.
[100] O.P. Kapoor and S.B. Mathur, *Some geometrical characterizations of inner product spaces*, Bull. Austral. Math. Soc. 24 No. 2 (1981), 239-246. MR 83d: 46028.
[101] T.A. Kilgore, *A Characterization of Lagrange Interpolating Projection with Minimal Tchebycheff Norm*, J. Approx. Th. 24, (1978), 273-288.
[102] T.A. Kilgore, *Optimal Interpolation with Incomplete Polynomials*, J. Approx. Th. 41, (1984), 279-290.
[103] S. Kinnunen, *On projections and Birkhoff-James orthogonality in Banach spaces*, Nieuw. Archief. voor Wiskunde, No. 2 (1984), 235-255.
[104] J.A. Klarkson, *Uniformly convex spaces*, Trans. Amer. Math. Soc. 40 (1936), 396-414.
[105] A.L. Koldobskii, *Extention of isometries in Orlicz spaces*, (Russian) in: 30th Hercen reading Mathematics, Leningr. Gos. Pedag. Inst., Leningrad, 1977.
[106] A.L. Koldobskii, *Uniqueness theorems for measures in Banach spaces and their applications*, (Russian) Candidate Thesis, Leningr. Univers, Leningrad, 1982.
[107] A.L. Koldobskii and V.P. Odinec (= W.P. Odyniec), *On minimal projections generated by isometries of Banach spaces*, Comment. Math., 27, No. 2 (1989), 265-274.
[108] W.M. Kozłowski, *Modular Function Spaces*, Series of Monographs and Textbooks in Pure and Applied Mathematics, 122, Dekker, New York, Basel, 1988.
[109] M.A. Krasnosielskij and Ja. Rutickij, *Convex functions and Orlicz Spaces*, Gröningen 1961.
[110] S.G. Krein, Yu.I. Petunin and and E.M. Semenov, *Interpolation of linear operators*, (Russian) Nauka, Moscow, (1978), MR 81f:46086.
[111] H.E. Lacey, *The Isometric theory of Classical Banach Spaces*, Springer-Verlag, New York, 1974.
[112] J. Lamperti, *On the isometries of certain function spaces*, Pacif. J. Math., 8 No. 3 (1958), 459-466.
[113] A.Yu. Levin and Yu.I. Petunin, *Some problems related to the concept of orthogonality in a Banach space*, (Russian) Uspehi Matem. Nauk, 18, No. 3 (1963), 167-171. MR 27 #2833.
[114] S. Lewanowicz, *Minimal projective operators*, (Polish) Ann. Soc. Math. Polon. Ser III, Mat. Stosow., 15 (1979) 25-46. MR 81d:41040.
[115] G. Lewicki, *An existence theorem for minimal linear projections*, in: Constructive Theory of Functions 84, Sofia, 1984, 544-548.
[116] G. Lewicki, *A theorem of Bernstein's type for linear projections*, Univ. Iagellon. Acta Math. 27 (1988) 23-27.
[117] G. Lewicki, *Kolmogorov's type criteria for spaces of compact operators*, (to appear in J.Approx. Th.).

[118]. A. Lima, G. Olsen, *Extreme points in duals of complex operator space*, Proc Amer. Math. Soc. vol. 94, No. 3 (1988), 437-440.

[119] J. Lindenstrauss, *On the extention of operators with range in a C(K) space*, Proc. Amer. Math. Soc., 15 No. 2 (1964), 218-224. MR 29 #5089.

[120] J. Lindenstrauss, *On projections with norm one-an example*, Proc. Amer. Math. Soc. 15 No. 3 (1964), 403-406. MR 28 #4335.

[121] J. Lindenstrauss, *On nonseparable reflexive Banach spaces*, Bull. Amer. Math. Soc., 72 (1966), 967-970. MR 34 #4875.

[122] J. Lindenstrauss, *On the extention of operators with finite dimensional range*, Illinois J. Math. No. 8 (1964), 488-499. MR 29 #6317.

[123] J. Lindenstrauss and A. Pełczyński, *Contributions to the theory of the Classical Banach spaces*, J. of Functional Analysis, 8 No. 2 (1971), 225-249. MR 45 #863.

[124] J. Lindenstrauss and L. Tzafriri, *Classical Banach Spaces*, Lecture Notes in Mathematics, vol. 338, Springer-Verlag, Berlin 1973.

[125] Yu.L. Ljubič, *Conservative operators*, (Russian) Uspehi Matem. Nauk, 20 No. 2 (1965), 221-225. MR 34 #4925.

[126] V.V. Lokot, *The norm of projective operators in the space* l_1^n, (Russian), in: Application of functional analysis in approximation theory, Kalinin University, Kalinin, 1978, 108-115.

[127] V.V. Lokot, M.B. Anohin, N.I. Komleva and N.B. Cvetkova, *Uniqueness of projection operators with the minimal norms onto hyperspaces in the space* l_1^n, Deposed manuscript, DEP - 5369-83. RJM #84 1B 923.

[128] V.V. Lokot, *The problem of unicity of minimal projections onto two dimensional subspace in* l_1^4, WINITI No. 4696 (1988) (Russian).

[129] J. Łoś, *Mathematical Theory of von Neumann Economic Models*, Colloq. Math. Prace Mat. 40 No. 2 (1979), 327-346.

[130] S.M. Lozinskii, *On a class of linear operators*, (Russian) Dokl. Acad. Nauk SSSR, 61 No. 2 (1948), 193-196. MR 10-188.

[131] W. Luski, *A note of rotations in separable Banach spaces*, Studia Math., 65 No. 3 (1979), 239-242. MR 81c:40010.

[132] V.L. Makarov and A.M. Rubinov, *A mathematical theory of economics dynamices and equilibrium*, (Russian), Nauka, Moscow, 1973, MR 51 #9766.

[133] J.C. Mason, *Minimal projections and near-best approximations by multivariate polynomial expansion and interpolation*, in: Multivariate Approximation Theory II.

[134] P. Mc Mullen, *On zonotopes*, Trans. Amer. Math. Soc. 159 (1971), 91-109. MR 43 #5410.

[135] V.D. Milman, *The geometric theory of Banach spaces Part II. (The geometry of unit sphere)*, (Russian) Uspehi Matem. Nauk, 26 No. 6 (1971), 78-149.

[136] J. Musielak, *Orlicz Spaces and Modular Spaces*, Lecture Notes in Math., vol. 1034, Springer-verlag, Berlin, Heidelberg, New York, Tokyo (1988).

[137] D.J. Newman and H.S. Shapiro, *Some theorems on Chebyshev approximation*, Duke Math. J. 30 (1963), 673-681.

[138] Nguyen To Nhu, *Uniform retracts and extentions of uniformly continuous maps, Lipschitz maps and metrics in uniform and metric spaces*, Thesis, Warsaw University, Warsaw, 1979.

[139] V.P. Odinec (= W.P. Odyniec), *On projections with unit norm and related problems of geometry of Banach spaces*, (Russian), Candidate thesis, Leningr. Gos. Pedag. Inst., Leningrad, 1975.

[140] V.P. Odinec, *The Gâteaux differential and uniqueness of norm preserving extention of linear operators*, (Russian) Izv. Vyss. Učebn. Zaved. Matematika, No. 4 (1973), 77-86. MR 49 #1108.

[141] V.P. Odinec, *The uniqueness of projection with norm equal to one in a Banach space*, (Russian) Izv. Vyss. Učebn. Zaved. Matematika, No. 1 (1974), 82-89. MR 50 #2876.

[142] V.P. Odinec, *On uniqueness of minimal projections in Banach space*, (Russian) Dokl. Akad. Nauk SSSR, 220 No. 4 (1975), 779-781. = {Soviet Math. Dokl. 16 No. 1 (1975), 151-154. (English translation)}. MR 57 #10412.
[143] V.P. Odinec, *Smoothness subspaces, normal bases and the uniqueness of projections with norm one*, (Russian) Rev. Roumain Math. Pur. Appl. 20 No. 4 (1975), 429-437. MR 58 #7042.
[144] V.P. Odinec, *Conditions for uniqueness of a projection with unit norm*, (Russian), Mat. Zamietki 22 No. 6 (1977), 45-49, 928 = {Math. Notes 22 (1978), 515-517. (English translation)}. MR 56 #12847.
[145] V.P. Odinec, *The uniqueness of minimal projection*, (Russian) Izv. Vyss. Učebn. Zaved. Matematika, No. 3 (1978), 73-75 = {Soviet Math. (Iz. VUZ), 22 No. 2 (1978), 64-66 (English translation)}. MR 58 #30072.
[146] V.P. Odinec, *Minimal projections. Uniqueness conditions*, (Russian) Teor. Funkcii, Functional. Anal. i Priloz. Vyp. 30 (1978), 101-108. MR 80 c:46029.
[147] V.P. Odinec, *Elementary rotations in Banach spaces*, (Russian) Deposed manuscript, DEP-2755-79 (Adnotation in Izv. Vyss. Učebn. Zaved. Matematika, No. 10 (1979), 103-104). ZBl 422 #46038.
[148] V.P. Odinec, *On the uniqueness of minimal projections in* l^n_∞ $(n\geq 3)$, Bull. Acad. Polon. Sci., Ser. math., 28 No. 7-8 (1980), 347-350. MR. 83d:46020.
[149] V.P. Odinec, *Remarks on the uniqueness of minimal projections with nonunit norm*, Bull. Acad. Polon. Sci. Ser. math. 29 No. 3-4 (1981), 145-151. MR 82m:46017.
[150] V.P. Odinec, *Conditions for the existence and strong uniqueness of a projection with unit norm*, (Russian) Mat. Zamietki, 32 No. 5 (1982), 607-612 = {Math. Notes 32 No.5-6 (1982), 788-791 (English translation)}. MR 85f:46031.
[151] V.P. Odinec, *On a property of reflexive Banach spaces with transitive norm*, Bull. Acad. Polon. Sci., Ser. math. 30 No. 7-8 (1982), 353-357. MR 84i: 46031.
[152] V.P. Odinec, *Strong uniqueness of minimal projections in Banach spaces*, (Russian) Izv. Vyss. Učebn. Zaved. Matematika, No. 9 (1984), 75-77 ={Soviet Math. (Iz. VUZ) 28 (1984), 105-108 (English translation)}. MR 86g:46026.
[153] V.P. Odinec, *On the uniqueness of minimal projection in* l_1, in: Problemy matematyczne, 1985, Zeszyt 7, WSP, Bydgoszcz, 1986, 5-10.
[154] V.P Odinec, *Functionals dually generated by minimal projections and criteria for Banach spaces to be Hilbert spaces*, (Russian), Mat. Zamietki 38 No. 5 (1985), 770-776 = {Math. Notes 38 (1985) (English translation)}.
[155] V.P. Odinec, *Codimension one minimal projections in Banach spaces and a mathematical programming problem*, Dissertationes Math. Rozprawy matematyczne. vol. 254, 1986.
[156] V.P. Odinec, *Minimal projections in Banach spaces. Problems of existence and uniqueness and theirs applications*, (Russian), Bydgoszcz, 1985.
[157] P.L. Papini, *Approximation and norms derivatives in real normed spaces*, Resultate Mathem., 5 No. 1 (1982), 81-94. MR 840:41044.
[158] T. Parthasarathy and T.E.S. Raghavan, *Some topics in two person games*, Modern Analytic and Comput. Methods in Science and Math., No.22 Amer. Elsevier, New York, 1971, MR 43 #2996.
[159] A. Pełczyński, *Projection in certain Banach spaces*, Studia Math., 19 (1960), 209-228. MR 23 #A3441.
[160] A. Pełczyński, *Linear extentions, linear averagings and their applications to linear topological classification of spaces of continuous functions*, Dissertationes Math. Rozprawy Mat., vol. 58, 1968. MR 37 #3335.
[161] A. Pełczyński, *Certain problems of Banach*, (Russian) Uspehi Matem. Nauk, 28 No. 6 (1973), 67-75. MR 57 #1087.

[162] R.R. Phelps, *Uniqueness of Hahn-Banach extentions and unique best approximation*, Trans. Amer. Math. Soc. 95 (1960), 238-255. MR 22 #3964.

[163] A.I. Plotkin, *Isometric operators on subspaces of* L_p, (Russian) Dokl. Acad. Nauk SSSR, 193 No. 3 (1970), 537-539. MR 42 #6601.

[164] E.T. Poulsen, *Eindeutige Hahn-Banach Erweiterungen*, Math. Annalen, 162 (1966), 225-227. MR 33 #6340.

[165] M Riesz, *Sur les maksima des formes bilinéares et sur les fonctionelles linéaires*, Acta Math. 49 (1926), 495-497.

[166] R.T. Rockafellar, *Convex analysis*, Princeton University Press, Princeton, New Jersey, 1970.

[167] S. Rolewicz, *Metric linear spaces*, Polish Scientific Publishers and D. Reidel Publishing Company, Warsaw 1985.

[168] S. Rolewicz, *On minimal projections of the space* $L^p([0,1])$ *on one-codimensional subspace*, Bull. Acad. Polon. Sci. Math. 34 No.3-4 (1986), 151-153.

[169] S. Rolewicz, *On projections on spaces of finite codimension in Orlicz spaces, Of infimum of norm of projections on subspaces codimension one, On projections on subspaces of finite codimension*; Polish Academy of Sciences, preprint No. 436 (1988).

[170] I.V. Romanowskii, *Algorithms for the solution of extremel problems*, (Russian) Nauka, Moscow, 1977. MR 58 #4294.

[171] H.H. Schaefer, *Topological vector spaces*, Springer-Verlag, New York, 1971.

[172] R. Schumaker, *On the uniqueness property of minimal projections*, J. Approx. Th. 31 (1981), 107-117.

[173] E.M. Semenov, *Embedding theorems for Banach spaces of measurable functions*, (Russian), Dokl. Acad. Nauk SSSR, 156 No. 6 (1964), 1292-1295. MR 30 #3368.

[174] E.M. Semenov, *Interpolation of linear operators in symmetric spaces*, Doktor. Thesis, Voronezsk. Gos. Univers. Vorenez, 1968.

[175] I. Singer, *On Banach spaces with symmetric bases*, (Russian) Rev. Romain. Math. Pur. Appl. 6 No. 1 (1961), 169-176. MR 26 #4152.

[176] I. Singer, *Some characterizations of symmetric bases in Banach spaces*, Bull. Acad. Polon. Sci., Ser. mat. 10 No. 4 (1962), 185-192.

[177] I. Singer, *Bases in Banach spaces, Part I*, Springer-Verlag, Berlin-New York, 1970.

[178] I. Singer, *Bases in Banach spaces, Part II*, Springer-Verlag, Berlin-New York, 1980.

[179] A.I. Skorik, *Isometries of ideal coordinate spaces*, (Russian) Uspehi Mat. Nauk, 31 No. 2 (1976), 229-230.

[180] A. Sobczyk, *Projections of the space (m) on its subspace* (c_o), Bull. Amer. Math. Soc., 47 (1941), 938-947.

[181] A. Sobczyk, *Projections in Minkowski and Banach spaces*, Duke Math. J. 8 (1941), 78-106.

[182] M.Z. Solomjak, *On orthogonal bases in Banach spaces*, (Russian) Vestnik Leningr. Gos. Univ., Ser. Mat. No. 1 (1957), 27-36. MR 19-45.

[183] V.N. Sudakov, *Geometric problems of the theory of infinite dimensional probability distributions*, (Russian) Trudy Mat. Inst. Steklov, vol. 141, Leningrad, 1976. MR 55 #4359.

[184] J. Sudolski and A. Wójcik, *Some remarks on strong uniqueness of best approximation*, Polish Academy of Sciences, preprint, No. 448 (1989).

[185] R. Sztencel *and P. Zaremba, On self-conjugate Banach spaces*, Coloq. Math. 44 No. 1 (1981), 111-115. MR 83b:46013.

[186] A.E. Taylor, *The extention of linear functionals*, Duke Math. J. 5 (1936), 538-547.

[187] L.P. Vlasov, *Several theorems on Čebysev sets*, (Russian) Matem. Zamietki, 11 No. 2 (1972), 135-144. MR 45 #9046.

[188] W. Wojtyński, *Banach spaces in which the isometries act transitively on the unit sphere*, Bull. Acad. Polon. Sci., Ser. math., 22 No. 8 (1974), 925-929.

[189] A. Wójcik, *Characterizations of strong unicity by tangent cones*, in: Approximation and function spaces, Proc of the international conference held in Gdańsk, August 27-31, 1979, ed. by Z. Ciesielski, PWN Warszawa, North-Holland, Amsterdam-New York-Oxford, (1981), 854-866.

[190] D.E. Wulbert, *Some complemented function spaces in C(X)*, Pacif. J. Math., 24 No. 3 (1968), 589-602. MR 36 #6915.

[191] M.G. Zaidenberg and A.I. Skorik, *Groups of isometries that contain reflections*, (Russian) Functional. Anal. i Prilozen. 10 No. 4 (1976), 87-88 = (Functional Anal. Appl. 10 No. 4 (1976), 322-323 (1977), (English translation)). MR 55 #6240.

[192] S.I. Zuhovicki, *On the approximation of real functions in the sense of P.L. Chebyshev*, (Russian) Uspehi Mat. Nauk, 11 No. 2, 125-159. MR 19-30.

[193] H. Fürstenberg, *Ergodicity and Transformation of the Torus*, Amer. J. Math. 83, (1961), 573-601.

Author Index

Subject Index

Symbol Index